Statistics
FOR
DUMMIES®
2ND EDITION

by Deborah J. Rumsey, PhD

WILEY

Wiley Publishing, Inc.

Statistics For Dummies®, 2nd Edition

Published by
Wiley Publishing, Inc.
111 River St.
Hoboken, NJ 07030-5774
www.wiley.com

Copyright © 2011 by Wiley Publishing, Inc., Indianapolis, Indiana

Published simultaneously in Canada

For general information on our other products and services, please contact our Customer Care Department within the U.S. at 877-762-2974, outside the U.S. at 317-572-3993, or fax 317-572-4002.

For technical support, please visit www.wiley.com/techsupport.

Wiley also publishes its books in a variety of electronic formats. Some content that appears in print may not be available in electronic books.

Library of Congress Control Number: 2011921775

ISBN: 978-0-470-91108-2

ISBN 978-0-470-91108-2 (pbk); ISBN 978-1-118-01204-8 (ebk); ISBN 978-1-118-01205-5 (ebk); ISBN 978-1-118-01206-2 (ebk)

Manufactured in the United States of America

10 9

WILEY

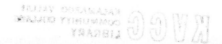

About the Author

Deborah J. Rumsey, PhD, is a Statistics Education Specialist and Auxiliary Professor in the Department of Statistics at The Ohio State University. Dr. Rumsey is a Fellow of the American Statistical Association. She has won the Presidential Teaching Award from Kansas State University and has been inducted into the Wall of Inspiration at her high school alma mater, Burlington High School, in Burlington, Wisconsin. She is also the author of *Statistics II For Dummies, Statistics Workbook For Dummies, Probability For Dummies,* and *Statistics Essentials For Dummies.* She has published numerous papers and given many professional presentations and workshops on the subject of statistics education. She is the original conference designer of the biennial United States Conference on Teaching Statistics (USCOTS). Her passions include being with her family, camping and bird watching, getting seat time on her Kubota tractor, and cheering the Ohio State Buckeyes on to their next national championship.

Dedication

To my husband Eric: My sun rises and sets with you. To my son Clint: I love you up to the moon and back.

Author's Acknowledgments

My heartfelt thanks to Lindsay Lefevere and Kathy Cox for the opportunity to write *For Dummies* books for Wiley; to my project editors Georgette Beatty, Corbin Collins, and Tere Drenth for their unwavering support and vision; to Marjorie Bond, Monmouth College, for agreeing to be my technical editor (again!); to Paul Stephenson, who also provided technical editing; and to Caitie Copple and Janet Dunn for great copy editing.

Special thanks to Elizabeth Stasny, Joan Garfield, Kythrie Silva, Kit Kilen, Peg Steigerwald, Mike O'Leary, Tony Barkauskas, Ken Berk, and Jim Higgins for inspiration and support along the way; and to my entire family for their steadfast love and encouragement.

Publisher's Acknowledgments

We're proud of this book; please send us your comments at http://dummies.custhelp.com. For other comments, please contact our Customer Care Department within the U.S. at 877-762-2974, outside the U.S. at 317-572-3993, or fax 317-572-4002.

Some of the people who helped bring this book to market include the following:

Acquisitions, Editorial, and Media Development

Project Editor: Corbin Collins

(Previous Edition: Tere Drenth)

Senior Project Editor: Georgette Beatty

Executive Editor: Lindsay Sandman Lefevere

Copy Editor: Caitlin Copple

(Previous Edition: Janet S. Dunn, PhD)

Assistant Editor: David Lutton

Technical Editors: Marjorie E. Bond, Paul L. Stephenson III

Editorial Manager: Michelle Hacker

Editorial Supervisor and Reprint Editor: Carmen Krikorian

Editorial Assistant: Jennette ElNaggar

Cover Photo: © iStockphoto.com/Norebbo

Cartoons: Rich Tennant (www.the5thwave.com)

Composition Services

Project Coordinator: Sheree Montgomery

Layout and Graphics: Carrie A. Cesavice, Corrie Socolovitch

Proofreaders: Dwight Ramsey, Shannon Ramsey

Indexer: Christine Karpeles

Publishing and Editorial for Consumer Dummies

Kathleen Nebenhaus, Vice President and Executive Publisher

Kristin Ferguson-Wagstaffe, Product Development Director

Ensley Eikenburg, Associate Publisher, Travel

Kelly Regan, Editorial Director, Travel

Publishing for Technology Dummies

Andy Cummings, Vice President and Publisher

Composition Services

Debbie Stailey, Director of Composition Services

Contents at a Glance

Table of Contents

Part V: Statistical Studies and the Hunt for a Meaningful Relationship 243

Chapter 16: Polls, Polls, and More Polls 245

Chapter 17: Experiments: Medical Breakthroughs or Misleading Results? 261

Chapter 18: Looking for Links: Correlation and Regression 279

Introduction

• •

*Y*ou get hit with an incredible amount of statistical information on a daily basis. You know what I'm talking about: charts, graphs, tables, and headlines that talk about the results of the latest poll, survey, experiment, or other scientific study. The purpose of this book is to develop and sharpen your skills in sorting through, analyzing, and evaluating all that info, and to do so in a clear, fun, and pain-free way. You also gain the ability to decipher and make important decisions about statistical results (for example, the results of the latest medical studies), while being ever aware of the ways that people can mislead you with statistics. And you see how to do it right when it's your turn to design the study, collect the data, crunch the numbers, and/or draw the conclusions.

This book is also designed to help those of you out there who are taking an introductory statistics class and can use some back-up. You'll gain a working knowledge of the big ideas of statistics and gather a boatload of tools and tricks of the trade that'll help you get ahead of the curve when you take your exams.

This book is chock-full of real examples from real sources that are relevant to your everyday life — from the latest medical breakthroughs, crime studies, and population trends to the latest U.S. government reports. I even address a survey on the worst cars of the millennium! By reading this book, you'll understand how to collect, display, and analyze data correctly and effectively, and you'll be ready to critically examine and make informed decisions about the latest polls, surveys, experiments, and reports that bombard you every day. You even find out how to use crickets to gauge temperature!

You also get to enjoy poking a little fun at statisticians (who take themselves too seriously at times). After all, with the right skills and knowledge, you don't have to be a statistician to understand introductory statistics.

About This Book

This book departs from traditional statistics texts, references, supplemental books, and study guides in the following ways:

- ✔ It includes practical and intuitive explanations of statistical concepts, ideas, techniques, formulas, and calculations found in an introductory statistics course.
- ✔ It shows you clear and concise step-by-step procedures that explain how you can intuitively work through statistics problems.

- ✔ It includes interesting real-world examples relating to your everyday life and workplace.

- ✔ It gives you upfront and honest answers to your questions like, "What does this really mean?" and "When and how will I ever use this?"

Conventions Used in This Book

You should be aware of three conventions as you make your way through this book:

- ✔ **Definition of sample size (*n*):** When I refer to the size of a sample, I mean the final number of individuals who participated in and provided information for the study. In other words, *n* stands for the size of the final data set.

- ✔ **Dual-use of the word *statistics*:** In some situations, I refer to statistics as a subject of study or as a field of research, so the word is a singular noun. For example, "Statistics is really quite an interesting subject." In other situations, I refer to statistics as the plural of *statistic,* in a numerical sense. For example, "The most common statistics are the mean and the standard deviation."

- ✔ **Use of the word *data*:** You're probably unaware of the debate raging amongst statisticians about whether the word *data* should be singular ("data is . . .") or plural ("data are . . ."). It got so bad that recently one group of statisticians had to develop two different versions of a statistics T-shirt: "Messy Data Happens" and "Messy Data Happen." At the risk of offending some of my colleagues, I go with the plural version of the word *data* in this book.

- ✔ **Use of the term *standard deviation*:** When I use the term *standard deviation,* I mean *s,* the sample standard deviation. (When I refer to the population standard deviation, I let you know.)

Here are a few other basic conventions to help you navigate this book:

- ✔ I use *italics* to let you know a new statistical term is appearing on the scene.

- ✔ If you see a **boldfaced** term or phrase in a bulleted list, it's been designated as a keyword or key phrase.

- ✔ Addresses for Web sites appear in `monofont`.

What You're Not to Read

I like to think that you won't skip anything in this book, but I also know you're a busy person. So to save time, feel free to skip anything marked with the

Technical Stuff icon as well as text in sidebars (the shaded gray boxes that appear throughout the book). These items feature information that's interesting but not crucial to your basic knowledge of statistics.

Foolish Assumptions

I don't assume that you've had any previous experience with statistics, other than the fact that you're a member of the general public who gets bombarded every day with statistics in the form of numbers, percents, charts, graphs, "statistically significant" results, "scientific" studies, polls, surveys, experiments, and so on.

What I do assume is that you can do some of the basic mathematical operations and understand some of the basic notation used in algebra, such as the variables x and y, summation signs, taking the square root, squaring a number, and so on. If you need to brush up on your algebra skills, check out *Algebra I For Dummies*, 2nd Edition, by Mary Jane Sterling (Wiley).

I don't want to mislead you: You do encounter formulas in this book, because statistics does involve a bit of number crunching. But don't let that worry you. I take you slowly and carefully through each step of any calculations you need to do. I also provide examples for you to work along with this book, so that you can become familiar and comfortable with the calculations and make them your own.

How This Book Is Organized

This book is organized into five parts that explore the major areas of introductory statistics, along with a final part that offers some quick top-ten nuggets for your information and enjoyment. Each part contains chapters that break down each major area of statistics into understandable pieces.

Part I: Vital Statistics about Statistics

This part helps you become aware of the quantity and quality of statistics you encounter in your workplace and your everyday life. You find out that a great deal of that statistical information is incorrect, either by accident or by design. You take a first step toward becoming statistically savvy by recognizing some of the tools of the trade, developing an overview of statistics as a process for getting and interpreting information, and getting up to speed on some statistical jargon.

Part II: Number-Crunching Basics

This part helps you become more familiar and comfortable with making, interpreting, and evaluating data displays (otherwise known as charts, graphs, and so on) for different types of data. You also find out how to summarize and explore data by calculating and combining some commonly used statistics as well as some statistics you may not know about yet.

Part III: Distributions and the Central Limit Theorem

In this part, you get into all the details of the three most common statistical distributions: the binomial distribution, the normal (and standard normal, also known as Z-distribution), and the t-distribution. You discover the characteristics of each distribution and how to find and interpret probabilities, percentiles, means, and standard deviations. You also find measures of relative standing (like percentiles).

Finally, you discover how statisticians measure variability from sample to sample and why a measure of precision in your sample results is so important. And you get the lowdown on what some statisticians describe as the "Crowning Jewel of all Statistics": the Central Limit Theorem (CLT). I don't use quite this level of flourishing language to describe the CLT; I just tell my students it's an MDR ("Mighty Deep Result"; coined by my PhD adviser). As for how my students describe their feelings about the CLT, I'll leave that to your imagination.

Part IV: Guesstimating and Hypothesizing with Confidence

This part focuses on the two methods for taking the results from a sample and generalizing them to make conclusions about an entire population. (Statisticians call this process *statistical inference.*) These two methods are confidence intervals and hypothesis tests.

In this part, you use confidence intervals to come up with good estimates for one or two population means or proportions, or for the difference between them (for example, the average number of hours teenagers spend watching TV per week or the percentage of men versus women in the United States who take arthritis medicine every day). You get the nitty-gritty on how confidence intervals are formed, interpreted, and evaluated for correctness and credibility. You explore the factors that influence the width of a confidence interval (such as sample size) and work through formulas, step-by-step calculations, and examples for the most commonly used confidence intervals.

The hypothesis tests in this part show you how to use your data to test someone's claim about one or two population means or proportions, or the difference between them. (For example, a company claims their packages are delivered in two days on average — is this true?) You discover how researchers (should) go about forming and testing hypotheses and how you can evaluate their results for accuracy and credibility. You also get detailed step-by-step directions and examples for carrying out and interpreting the results of the most commonly used hypothesis tests.

Part V: Statistical Studies and the Hunt for a Meaningful Relationship

This part gives an overview of surveys, experiments, and observational studies. You find out what these studies do, how they are conducted, what their limitations are, and how to evaluate them to determine whether you should believe the results.

You also get all the details on how to examine pairs of numerical variables and categorical variables to look for relationships; this is the object of a great number of studies. For pairs of categorical variables, you create two-way tables and find joint, conditional, and marginal probabilities and distributions. You check for independence, and if a dependent relationship is found, you describe the nature of the relationship using probabilities. For numerical variables you create scatterplots, find and interpret correlation, perform regression analyses, study the fit of the regression line and the impact of outliers, describe the relationship using the slope, and use the line to make predictions. All in a day's work!

Part VI: The Part of Tens

This quick and easy part shares ten ways to be a statistically savvy sleuth and root out suspicious studies and results, as well as ten surefire ways to boost your statistics exam score.

Some statistical calculations involve the use of statistical tables, and I provide quick and easy access to all the tables you need for this book in the appendix. These tables are the Z-table (for the standard normal, also called the Z-distribution), the t-table (for the t-distribution), and the binomial table (for — you guessed it — the binomial distribution). Instructions and examples for using these three tables are provided in their corresponding sections of this book.

Icons Used in This Book

Icons are used in this book to draw your attention to certain features that occur on a regular basis. Here's what they mean:

This icon refers to helpful hints, ideas, or shortcuts that you can use to save time. It also highlights alternative ways to think about a particular concept.

This icon is reserved for particular ideas that I hope you'll remember long after you read this book.

This icon refers to specific ways that researchers or the media can mislead you with statistics and tells you what you can do about it. It also points out potential problems and cautions to keep an eye out for on exams.

This icon is a sure bet if you have a special interest in understanding the more technical aspects of statistical issues. You can skip this icon if you don't want to get into the gory details.

Where to Go from Here

This book is written in such a way that you can start anywhere and still be able to understand what's going on. So you can take a peek at the table of contents or the index, look up the information that interests you, and flip to the page listed. However if you have a specific topic in mind and are eager to dive into it, here are some directions:

- ✔ To work on finding and interpreting graphs, charts, means or medians, and the like, head to Part II.
- ✔ To find info on the normal, Z-, t-, or binomial distributions or the Central Limit Theorem, see Part III.
- ✔ To focus on confidence intervals and hypothesis tests of all shapes and sizes, flip to Part IV.
- ✔ To delve into surveys, experiments, regression, and two-way tables, see Part V.

Or if you aren't sure where you want to start, you may just go with Chapter 1 for the big picture and then plow your way through the rest of the book. Happy reading!

Part I
Vital Statistics about Statistics

The 5th Wave   By Rich Tennant

"Is it just me or did the whole '50% satisfaction' statistic seem a little unimpressive?"

In this part . . .

When you turn on the TV or open a newspaper, you're bombarded with numbers, charts, graphs, and statistical results. From today's poll to the latest major medical breakthroughs, the numbers just keep coming. Yet much of the statistical information you're asked to consume is actually wrong — by accident or even by design. How is a person to know what to believe? By doing a lot of good detective work.

This part helps awaken the statistical sleuth that lies within you by exploring how statistics affect your everyday life and your job, how bad much of the information out there really is, and what you can do about it. This part also helps you get up to speed with some useful statistical jargon.

Chapter 1

Statistics in a Nutshell

In This Chapter

▶ Finding out what the process of statistics is all about

▶ Gaining success with statistics in your everyday life, your career, and in the classroom

The world today is overflowing with data to the point where anyone (even me!) can be overwhelmed. I wouldn't blame you if you were cynical right now about statistics you read about in the media — I am too at times. The good news is that while a great deal of misleading and incorrect information is lying out there waiting for you, a lot of great stuff is also being produced; for example, many studies and techniques involving data are helping improve the quality of our lives. Your job is to be able to sort out the good from the bad and be confident in your ability to do that. Through a strong understanding of statistics and statistical procedures, you gain power and confidence with numbers in your everyday life, in your job, and in the classroom. That's what this book is all about.

In this chapter, I give you an overview of the role statistics plays in today's data-packed society and what you can do to not only survive but thrive. You get a much broader view of statistics as a partner in the scientific method — designing effective studies, collecting good data, organizing and analyzing the information, interpreting the results, and making appropriate conclusions. (And you thought statistics was just number-crunching!)

Thriving in a Statistical World

It's hard to get a handle on the flood of statistics that affect your daily life in large and small ways. It begins the moment you wake up in the morning and check the news and listen to the meteorologist give you her predictions for the weather based on her statistical analyses of past data and present weather conditions. You pore over nutritional information on the side of your cereal box while you eat breakfast. At work you pull numbers from charts and tables, enter data into spreadsheets, run diagnostics, take measurements, perform calculations, estimate expenses, make decisions using statistical baselines, and order inventory based on past sales data.

At lunch you go to the No. 1 restaurant based on a survey of 500 people. You eat food that was priced based on marketing data. You go to your doctor's appointment where they take your blood pressure, temperature, weight, and do a blood test; after all the information is collected, you get a report showing your numbers and how you compare to the statistical norms.

You head home in your car that's been serviced by a computer running statistical diagnostics. When you get home, you turn on the news and hear the latest crime statistics, see how the stock market performed, and discover how many people visited the zoo last week.

At night, you brush your teeth with toothpaste that's been statistically proven to fight cavities, read a few pages of your *New York Times* Best-Seller (based on statistical sales estimates), and go to sleep — only to start it all over again the next morning. But how can you be sure that all those statistics you encounter and depend on each day are correct? In Chapter 2, I discuss in more depth a few examples of how statistics is involved in our lives and workplaces, what its impact is, and how you can raise your awareness of it.

Some statistics are vague, inappropriate, or just plain wrong. You need to become more aware of the statistics you encounter each day and train your mind to stop and say "wait a minute!", sift through the information, ask questions, and raise red flags when something's not quite right. In Chapter 3, you see ways in which you can be misled by bad statistics and develop skills to think critically and identify problems before automatically believing results.

Like any other field, statistics has its own set of jargon, and I outline and explain some of the most commonly used statistical terms in Chapter 4. Knowing the language increases your ability to understand and communicate statistics at a higher level without being intimidated. It raises your credibility when you use precise terms to describe what's wrong with a statistical result (and why). And your presentations involving statistical tables, graphs, charts, and analyses will be informational and effective. (Heck, if nothing else, you need the jargon because I use it throughout this book; don't worry though, I always review it.)

In the next sections, you see how statistics is involved in each phase of the scientific method.

Designing Appropriate Studies

Everyone's asking questions, from drug companies to biologists; from marketing analysts to the U.S. government. And ultimately, everyone will use statistics to help them answer their questions. In particular, many medical and psychological studies are done because someone wants to know the answer to a question. For example,

- ✔ Will this vaccine be effective in preventing the flu?
- ✔ What do Americans think about the state of the economy?
- ✔ Does an increase in the use of social networking Web sites cause depression in teenagers?

The first step after a research question has been formed is to design an effective study to collect data that will help answer that question. This step amounts to figuring out what process you'll use to get the data you need. In this section, I give an overview of the two major types of studies — surveys and experiments — and explore why it's so important to evaluate how a study was designed before you believe the results.

Surveys

An *observational study* is one in which data is collected on individuals in a way that doesn't affect them. The most common observational study is the survey. *Surveys* are questionnaires that are presented to individuals who have been selected from a population of interest. Surveys take many different forms: paper surveys sent through the mail, questionnaires on Web sites, call-in polls conducted by TV networks, phone surveys, and so on.

If conducted properly, surveys can be very useful tools for getting information. However, if not conducted properly, surveys can result in bogus information. Some problems include improper wording of questions, which can be misleading, lack of response by people who were selected to participate, or failure to include an entire group of the population. These potential problems mean a survey has to be well thought out before it's given.

Many researchers spend a great deal of time and money to do good surveys, and you'll know (by the criteria I discuss in Chapter 16) that you can trust them. However, as you are besieged with so many different types of surveys found in the media, in the workplace, and in many of your classes, you need to be able to quickly examine and critique how a survey was designed and conducted and be able to point out specific problems in a well-informed way. The tools you need for sorting through surveys are found in Chapter 16.

Experiments

An *experiment* imposes one or more treatments on the participants in such a way that clear comparisons can be made. After the treatments are applied, the responses are recorded. For example, to study the effect of drug dosage on blood pressure, one group may take 10 mg of the drug, and another group may take 20 mg. Typically, a control group is also involved, in which subjects each receive a fake treatment (a sugar pill, for example), or a standard, non-experimental treatment (like the existing drugs given to AIDS patients.)

Good and credible experiments are designed to minimize bias, collect lots of good data, and make appropriate comparisons (treatment group versus control group). Some potential problems that occur with experiments include researchers and/or subjects who know which treatment they got, factors not controlled for in the study that affect the outcome (such as weight of the subject when studying drug dosage), or lack of a control group (leaving no baseline to compare the results with).

But when designed correctly, an experiment can help a researcher establish a cause-and-effect relationship if the difference in responses between the treatment group and the control group is statistically significant (unlikely to have occurred just by chance).

Experiments are credited with helping to create and test drugs, determining best practices for making and preparing foods, and evaluating whether a new treatment can cure a disease, or at least reduce its impact. Our quality of life has certainly been improved through the use of well-designed experiments. However, not all experiments are well-designed, and your ability to determine which results are credible and which results are incredible (pun intended) is critical, especially when the findings are very important to you. All the info you need to know about experiments and how to evaluate them is found in Chapter 17.

Collecting Quality Data

After a study has been designed, be it a survey or an experiment, the individuals who will participate have to be selected, and a process must be in place to collect the data. This phase of the process is critical to producing credible data in the end, and this section hits the highlights.

Selecting a good sample

Statisticians have a saying, "Garbage in equals garbage out." If you select your *subjects* (the individuals who will participate in your study) in a way that is *biased* — that is, favoring certain individuals or groups of individuals — then your results will also be biased. It's that simple.

Suppose Bob wants to know the opinions of people in your city regarding a proposed casino. Bob goes to the mall with his clipboard and asks people who walk by to give their opinions. What's wrong with that? Well, Bob is only going to get the opinions of a) people who shop at that mall; b) on that particular day; c) at that particular time; d) and who take the time to respond.

Those circumstances are too restrictive — those folks don't represent a cross section of the city. Similarly, Bob could put up a Web site survey and ask people to use it to vote. However, only people who know about the site, have Internet access, and want to respond will give him data, and typically only those with strong opinions will go to such trouble. In the end, all Bob has is a bunch of biased data on individuals that don't represent the city at all.

To minimize bias in a survey, the key word is *random*. You need to select your sample of individuals *randomly* — that is, with some type of "draw names out of a hat" process. Scientists use a variety of methods to select individuals at random, and you see how they do it in Chapter 16.

Note that in designing an experiment, collecting a random sample of people and asking them to participate often isn't ethical because experiments impose a treatment on the subjects. What you do is send out requests for volunteers to come to you. Then you make sure the volunteers you select from the group represent the population of interest and that the data is well collected on those individuals so the results can be projected to a larger group. You see how that's done in Chapter 17.

After going through Chapters 16 and 17, you'll know how to dig down and analyze others' methods for selecting samples and even be able to design a plan you can use to select a sample. In the end, you'll know when to say "Garbage in equals garbage out."

Avoiding bias in your data

Bias is the systematic favoritism of certain individuals or certain responses. Bias is the nemesis of statisticians, and they do everything they can to minimize it. Want an example of bias? Say you're conducting a phone survey on job satisfaction of Americans; if you call people at home during the day between 9 a.m. and 5 p.m., you miss out on everyone who works during the day. Maybe day workers are more satisfied than night workers.

You have to watch for bias when collecting survey data. For instance: Some surveys are too long — what if someone stops answering questions halfway through? Or what if they give you misinformation and tell you they make $100,000 a year instead of $45,000? What if they give you answers that aren't on your list of possible answers? A host of problems can occur when collecting survey data, and you need to be able to pinpoint those problems.

Experiments are sometimes even more challenging when it comes to bias and collecting data. Suppose you want to test blood pressure; what if the instrument you're using breaks during the experiment? What if someone quits the experiment halfway through? What if something happens during the experiment to distract the subjects or the researchers? Or they can't find a vein when they have to do a blood test exactly one hour after a dose of a drug is given? These problems are just some examples of what can go wrong in data collection for experiments, and you have to be ready to look for and find these problems.

After you go through Chapter 16 (on samples and surveys) and Chapter 17 (on experiments), you'll be able to select samples and collect data in an unbiased way, being sensitive to little things that can really influence the results. And you'll have the ability to evaluate the credibility of statistical results and to be heard, because you'll know what you're talking about.

Creating Effective Summaries

After good data have been collected, the next step is to summarize them to get a handle on the big picture. Statisticians describe data in two major ways: with numbers (called *descriptive statistics*) and with pictures (that is, charts and graphs).

Descriptive statistics

Descriptive statistics are numbers that describe a data set in terms of its important features:

✔ If the data are *categorical* (where individuals are placed into groups, such as gender or political affiliation), they are typically summarized using the number of individuals in each group (called the *frequency*) or the percentage of individuals in each group (called the *relative frequency*).

✔ *Numerical data* represent measurements or counts, where the actual numbers have meaning (such as height and weight). With numerical data, more features can be summarized besides the number or percentage in each group. Some of these features include

- Measures of center (in other words, where is the "middle" of the data?)

- Measures of spread (how diverse or how concentrated are the data around the center?)

- If appropriate, numbers that measure the relationship between two variables (such as height and weight)

Some descriptive statistics are more appropriate than others in certain situations; for example, the average isn't always the best measure of the center of a data set; the median is often a better choice. And the standard deviation isn't the only measure of variability on the block; the interquartile range has excellent qualities too. You need to be able to discern, interpret, and evaluate the types of descriptive statistics being presented to you on a daily basis and to know when a more appropriate statistic is in order.

The descriptive statistics you see most often are calculated, interpreted, compared, and evaluated in Chapter 5. These commonly used descriptive statistics include frequencies and relative frequencies (counts and percents) for categorical data; and the mean, median, standard deviation, percentiles, and their combinations for numerical data.

Charts and graphs

Data is summarized in a visual way using charts and/or graphs. These are displays that are organized to give you a big picture of the data in a flash and/or to zoom in on a particular result that was found. In this world of quick information and mini-sound bites, graphs and charts are commonplace. Most graphs and charts make their points clearly, effectively, and fairly; however, they can leave room for too much poetic license, and as a result, can expose you to a high number of misleading and incorrect graphs and charts.

In Chapters 6 and 7, I cover the major types of graphs and charts used to summarize both categorical and numerical data (see the preceding section for more about these types of data). You see how to make them, what their purposes are, and how to interpret the results. I also show you lots of ways that graphs and charts can be made to be misleading and how you can quickly spot the problems. It's a matter of being able to say "Wait a minute here! That's not right!" and knowing why not. Here are some highlights:

✔ Some of the basic graphs used for categorical data include pie charts and bar graphs, which break down variables, such as gender or which applications are used on teens' cellphones. A bar graph, for example, may display opinions on an issue using five bars labeled in order from "Strongly Disagree" up through "Strongly Agree." Chapter 6 gives you all the important info on making, interpreting, and, most importantly, evaluating these charts and graphs for fairness. You may be surprised to see how much can go wrong with a simple bar chart.

✔ For numerical data such as height, weight, time, or amount, a different type of graph is needed. Graphs called histograms and boxplots are used to summarize numerical data, and they can be very informative, providing excellent on-the-spot information about a data set. But of course they also can be misleading, either by chance or even by design. (See Chapter 7 for the scoop.)

You're going to run across charts and graphs every day — you can open a newspaper and probably find several graphs without even looking hard. Having a statistician's magnifying glass to help you interpret the information is critical so that you can spot misleading graphs before you draw the wrong conclusions and possibly act on them. All the tools you need are ready for you in Chapter 6 (for categorical data) and Chapter 7 (for numerical data).

Determining Distributions

A *variable* is a characteristic that's being counted, measured, or categorized. Examples include gender, age, height, weight, or number of pets you own. A *distribution* is a listing of the possible values of a variable (or intervals of values), and how often (or at what density) they occur. For example, the distribution of gender at birth in the United States has been estimated at 52.4% male and 47.6% female.

Different types of distributions exist for different variables. The following three distributions are the most commonly occurring distributions in an introductory statistics course, and they have many applications in the real world:

✔ If a variable is counting the number of successes in a certain number of trials (such as the number of people who got well by taking a certain drug), it has a *binomial* distribution.

✔ If the variable takes on values that occur according to a "bell-shaped curve," such as national achievement test scores, then that variable has a *normal* distribution.

✔ If the variable is based on sample averages and you have limited data, such as in a test of only ten subjects to see if a weight-loss program works, the *t*-distribution may be in order.

When it comes to distributions, you need to know how to decide which distribution a particular variable has, how to find probabilities for it, and how to figure out what the long-term average and standard deviation of the outcomes would be. To get you squared away on these issues, I've got three chapters for you, one dedicated to each distribution: Chapter 8 is all about the binomial, Chapter 9 handles the normal, and Chapter 10 focuses on the *t*-distribution.

For those of you taking an introductory statistics course (or any statistics course, for that matter), you know that one of the most difficult topics to understand is sampling distributions and the Central Limit Theorem (these two things go hand in hand). Chapter 11 walks you through these topics step by step so you understand what a sampling distribution is, what it's used for, and how it provides the foundation for data analyses like hypothesis tests and confidence intervals (see the next section for more about analyzing data). When you understand the Central Limit Theorem, it actually helps you solve difficult problems more easily, and all the keys to this information are there for you in Chapter 11.

Performing Proper Analyses

After the data have been collected and described using numbers and pictures, then comes the fun part: navigating through that black box called the *statistical analysis*. If the study has been designed properly, the original questions can be answered using the appropriate analysis — the operative word here being *appropriate*.

Many types of analyses exist, and choosing the right analysis for the right situation is critical, as is interpreting results properly, being knowledgeable of the limitations, and being able to evaluate others' choice of analyses and the conclusions they make with them.

In this book, you get all the information and tools you need to analyze data using the most common methods in introductory statistics: confidence intervals, hypothesis tests, correlation and regression, and the analysis of two-way tables. This section gives you a basic overview of those methods.

Margin of error and confidence intervals

You often see statistics that try to estimate numbers pertaining to an entire population; in fact, you see them almost every day in the form of survey results. The media tells you what the average gas price is in the U.S., how Americans feel about the job the president is doing, or how many hours people spend on the Internet each week.

But no one can give you a single-number result and claim it's an accurate estimate of the entire population unless he collected data on every single member of the population. For example, you may hear that 60 percent of the American people support the president's approach to healthcare, but you know they didn't ask you, so how could they have asked everybody? And since they didn't ask everybody, you know that a one-number answer isn't going to cut it.

What's really happening is that data is collected on a sample from the population (for example, the Gallup Organization calls 2,500 people at random), the results from that sample are analyzed, and conclusions are made regarding the entire population (for example, all Americans) based on those sample results.

The bottom line is, sample results vary from sample to sample, and this amount of variability needs to be reported (but it often isn't). The statistic used to measure and report the level of precision in someone's sample results is called the *margin of error*. In this context, the word *error* doesn't mean a mistake was made; it just means that because you didn't sample the entire population, a gap will exist between your results and the actual value you are trying to estimate for the population.

For example, someone finds that 60% of the 1,200 people surveyed support the president's approach to healthcare and reports the results with a margin of error of plus or minus 2%. This final result, in which you present your findings as a range of likely values between 58% and 62%, is called a *confidence interval*.

Everyone is exposed to results including a margin of error and confidence intervals, and with today's data explosion, many people are also using them in the workplace. Be sure you know what factors affect margin of error (like sample size) and what the makings of a good confidence interval are and how to spot them. You should also be able to find your own confidence intervals when you need to.

In Chapter 12, you find out everything you need to know about the margin of error: All the components of it, what it does and doesn't measure, and how to calculate it for a number of situations. Chapter 13 takes you step by step through the formulas, calculations, and interpretations of confidence intervals for a population mean, population proportion, and the difference between two means and proportions.

Hypothesis tests

One main staple of research studies is called hypothesis testing. A *hypothesis test* is a technique for using data to validate or invalidate a claim about a population. For example, a politician may claim that 80% of the people in her state agree with her — is that really true? Or, a company may claim that they deliver pizzas in 30 minutes or less; is that really true? Medical researchers use hypothesis tests all the time to test whether or not a certain drug is effective, to compare a new drug to an existing drug in terms of its side effects, or to see which weight-loss program is most effective with a certain group of people.

The elements about a population that are most often tested are

- The population mean (Is the average delivery time of 30 minutes really true?)

- The population proportion (Is it true that 80% of the voters support this candidate, or is it less than that?)

- The difference in two population means or proportions (Is it true that the average weight loss on this new program is 10 pounds more than the most popular program? Or, is it true that this drug decreases blood pressure by 10% more than the current drug?)

Hypothesis tests are used in a host of areas that affect your everyday life, such as medical studies, advertisements, polling data, and virtually anywhere that comparisons are made based on averages or proportions. And in the workplace, hypothesis tests are used heavily in areas like marketing, where you want to determine whether a certain type of ad is effective or whether a certain group of individuals buys more or less of your product now compared to last year.

Often you only hear the conclusions of hypothesis tests (for example, this drug is significantly more effective and has fewer side effects than the drug you are using now); but you don't see the methods used to come to these conclusions. Chapter 14 goes through all the details and underpinnings of hypothesis tests so you can conduct and critique them with confidence. Chapter 15 cuts right to the chase of providing step-by-step instructions for setting up and carrying out hypothesis tests for a host of specific situations (one population mean, one population proportion, the difference of two population means, and so on).

After reading Chapters 14 and 15, you'll be much more empowered when you need to know things like which group you should be marketing a product to; which brand of tires will last the longest; whether a certain weight-loss program is effective; and bigger questions like which surgical procedure you should opt for.

Correlation, regression, and two-way tables

One of the most common goals of research is to find links between variables. For example,

- Which lifestyle behaviors increase or decrease the risk of cancer?
- What side effects are associated with this new drug?
- Can I lower my cholesterol by taking this new herbal supplement?
- Does spending a large amount of time on the Internet cause a person to gain weight?

Finding links between variables is what helps the medical world design better drugs and treatments, provides marketers with info on who is more likely to buy their products, and gives politicians information on which to build arguments for and against certain policies.

In the mega-business of looking for relationships between variables, you find an incredible number of statistical results — but can you tell what's correct and what's not? Many important decisions are made based on these studies, and it's important to know what standards need to be met in order to deem the results credible, especially when a cause-and-effect relationship is being reported.

Chapter 18 breaks down all the details and nuances of plotting data from two numerical variables (such as dosage level and blood pressure), finding and interpreting *correlation* (the strength and direction of the linear relationship between x and y), finding the equation of a line that best fits the data (and when doing so is appropriate), and how to use these results to make predictions for one variable based on another (called *regression*). You also gain tools for investigating when a line fits the data well and when it doesn't, and what conclusions you can make (and shouldn't make) in the situations where a line does fit.

I cover methods used to look for and describe links between two categorical variables (such as the number of doses taken per day and the presence or absence of nausea) in detail in Chapter 19. I also provide info on collecting and organizing data into *two-way tables* (where the possible values of one variable make up the rows and the possible values for the other variable make up the columns), interpreting the results, analyzing the data from two-way tables to look for relationships, and checking for independence. And, as I do throughout this book, I give you strategies for critically examining results of these kinds of analyses for credibility.

Drawing Credible Conclusions

 To perform statistical analyses, researchers use statistical software that depends on formulas. But formulas don't know whether they are being used properly, and they don't warn you when your results are incorrect. At the end of the day, computers can't tell you what the results mean; you have to figure it out. Throughout this book you see what kinds of conclusions you can and can't make after the analysis has been done. The following sections provide an introduction to drawing appropriate conclusions.

Reeling in overstated results

Some of the most common mistakes made in conclusions are overstating the results or generalizing the results to a larger group than was actually represented by the study. For example, a professor wants to know which Super Bowl commercials viewers liked best. She gathers 100 students from her class on Super Bowl Sunday and asks them to rate each commercial as it is shown. A top-five list is formed, and she concludes that all Super Bowl viewers liked those five commercials the best. But she really only knows which ones *her students* liked best — she didn't study any other groups, so she can't draw conclusions about all viewers.

Questioning claims of cause and effect

One situation in which conclusions cross the line is when researchers find that two variables are related (through an analysis such as regression; see the earlier section "Correlation, regression, and two-way tables" for more info) and then automatically leap to the conclusion that those two variables have a cause-and-effect relationship.

For example, suppose a researcher conducted a health survey and found that people who took vitamin C every day reported having fewer colds than people who didn't take vitamin C every day. Upon finding these results, she wrote a paper and gave a press release saying vitamin C prevents colds, using this data as evidence.

Now, while it may be true that vitamin C does prevent colds, this researcher's study can't claim that. Her study was observational, which means she didn't control for any other factors that could be related to both vitamin C and colds. For example, people who take vitamin C every day may be more health conscious overall, washing their hands more often, exercising more, and eating better foods; all these behaviors may be helpful in reducing colds.

Until you do a controlled experiment, you can't make a cause-and-effect conclusion based on relationships you find. (I discuss experiments in more detail earlier in this chapter.)

Becoming a Sleuth, Not a Skeptic

Statistics is about much more than numbers. To really "get" statistics, you need to understand how to make appropriate conclusions from studying data and be savvy enough to not believe everything you hear or read until you find out how the information came about, what was done with it, and how the conclusions were drawn. That's something I discuss throughout the book, but I really zoom in on it in Chapter 20, which gives you ten ways to be a statistically savvy sleuth by recognizing common mistakes made by researchers and the media.

 For you students out there, Chapter 21 brings good statistical practice into the exam setting and gives you tips on increasing your scores. Much of my advice is based on understanding the big picture as well as the details of tackling statistical problems and coming out a winner on the other side.

 Becoming skeptical or cynical about statistics is very easy, especially after finding out what's going on behind the scenes; don't let that happen to you. You can find lots of good information out there that can affect your life in a positive way. Find a good channel for your skepticism by setting two personal goals:

- ✔ To become a well-informed consumer of the statistical information you see every day
- ✔ To establish job security by being the statistics "go-to" person who knows when and how to help others and when to find a statistician

Through reading and using the information in this book, you'll be confident in knowing you can make good decisions about statistical results. You'll conduct your own statistical studies in a credible way. And you'll be ready to tackle your next office project, critically evaluate that annoying political ad, or ace your next exam!

Chapter 2

The Statistics of Everyday Life

oday's society is completely taken over by numbers. Numbers are everywhere you look, from billboards showing the on-time statistics for a particular airline, to sports shows discussing the Las Vegas odds for upcoming football games. The evening news is filled with stories focusing on crime rates, the expected life span of junk-food junkies, and the president's approval rating. On a normal day, you can run into 5, 10, or even 20 different statistics (with many more on election night). Just by reading a Sunday newspaper all the way through, you come across literally hundreds of statistics in reports, advertisements, and articles covering everything from soup (how much does an average person consume per year?) to nuts (almonds are known to have positive health effects — what about other types of nuts?).

In this chapter I discuss the statistics that often appear in your life and work and talk about how statistics are presented to the general public. After reading this chapter, you'll realize just how often the media hits you with numbers and how important it is to be able to unravel the meaning of those numbers. Like it or not, statistics are a big part of your life. So, if you can't beat 'em, join 'em. And if you don't want to join 'em, at least try to understand 'em.

Statistics and the Media: More Questions than Answers?

Open a newspaper and start looking for examples of articles and stories involving numbers. It doesn't take long before numbers begin to pile up. Readers are inundated with results of studies, announcements of breakthroughs, statistical reports, forecasts, projections, charts, graphs, and summaries. The extent to which statistics occur in the media is mind-boggling. You may not even be aware of how many times you're hit with numbers nowadays.

This section looks at just a few examples from one Sunday paper's worth of news that I read the other day. When you see how frequently statistics are reported in the news without providing all the information you need, you may find yourself getting nervous, wondering what you can and can't believe anymore. Relax! That's what this book is for — to help you sort out the good information from the bad (the chapters in Part II give you a great start on that).

Probing popcorn problems

The first article I came across that dealt with numbers was "Popcorn plant faces health probe," with the subheading: "Sick workers say flavoring chemicals caused lung problems." The article describes how the Centers for Disease Control (CDC) expressed concern about a possible link between exposure to chemicals in microwave popcorn flavorings and some cases of fixed obstructive lung disease. Eight people from one popcorn factory alone contracted this lung disease, and four of them were awaiting lung transplants.

According to the article, similar cases were reported at other popcorn factories. Now, you may be wondering, what about the folks who eat microwave popcorn? According to the article, the CDC finds "no reason to believe that people who eat microwave popcorn have anything to fear." (Stay tuned.) The next step is to evaluate employees more in-depth, including conducting surveys to determine health and possible exposures to the said chemicals, checks of lung capacity, and detailed air samples. The question here is: How many cases of this lung disease constitute a real pattern, compared to mere chance or a statistical anomaly? (You find out more about this in Chapter 14.)

Venturing into viruses

The second article discussed a recent cyber attack: A wormlike virus made its way through the Internet, slowing down Web browsing and e-mail delivery across the world. How many computers were affected? The experts quoted in the article said that 39,000 computers were infected, and they in turn affected hundreds of thousands of other systems.

Questions: How did the experts get that number? Did they check each computer out there to see whether it was affected? The fact that the article was written less than 24 hours after the attack suggests the number is a guess. Then why say 39,000 and not 40,000 — to make it seem less like a guess? To find out more on how to guesstimate with confidence (and how to evaluate someone else's numbers), see Chapter 13.

Comprehending crashes

Next in the paper was an alert about the soaring number of motorcycle fatalities. Experts said that the *fatality rate* — the number of fatalities per 100,000 registered vehicles — for motorcyclists has been steadily increasing, as reported by the National Highway Traffic Safety Administration (NHTSA). In the article, many possible causes for the increased motorcycle death rate are discussed, including age, gender, size of engine, whether the driver had a license, alcohol use, and state helmet laws (or lack thereof). The report is very comprehensive, showing various tables and graphs with the following titles:

- Motorcyclists killed and injured, and fatality and injury rates by year, per number of registered vehicles, and per millions of vehicle miles traveled
- Motorcycle rider fatalities by state, helmet use, and blood alcohol content
- Occupant fatality rates by vehicle type (motorcycles, passenger cars, light trucks), per 10,000 registered vehicles and per 100 million vehicle miles traveled
- Motorcyclist fatalities by age group
- Motorcyclist fatalities by engine size (displacement)
- Previous driving records of drivers involved in fatal traffic crashes by type of vehicle (including previous crashes, DUI convictions, speeding convictions, and license suspensions and revocations)
- Percentage of alcohol-impaired motorcycle riders killed in traffic crashes by time of day, for single-vehicle, multiple-vehicle, and total crashes

This article is very informative and provides a wealth of detailed information regarding motorcycle fatalities and injuries in the U.S. However, the onslaught of so many tables, graphs, rates, numbers, and conclusions can be overwhelming and confusing and allow you to miss the big picture. With a little practice, and help from Part II, you'll be better able to sort out graphs, tables, and charts and all the statistics that go along with them. For example, some important statistical issues come up when you see rates versus counts (such as death rates versus number of deaths). As I address in Chapter 3, counts can give you misleading information if they're used when rates would be more appropriate.

Mulling malpractice

Further along in the newspaper was a report about a recent medical malpractice insurance study: Malpractice cases affect people in terms of the fees doctors charge and the ability to get the healthcare they need. The article indicates that 1 in 5 Georgia doctors have stopped doing risky procedures (such as delivering babies) because of the ever-increasing malpractice insurance rates in the state. This is described as a "national epidemic" and a "health crisis" around the country. Some brief details of the study are included, and the article states that of the 2,200 Georgia doctors surveyed, 2,800 of them — which they say represents about 18% of those sampled — were expected to stop providing high-risk procedures.

Wait a minute! That can't be right. Out of 2,200 doctors, 2,800 don't perform the procedures, and that is supposed to represent 18%? That's impossible! You can't have a bigger number on the top of a fraction, and still have the fraction be under 100%, right? This is one of many examples of errors in media reporting of statistics. So what's the real percentage? There's no way to tell from the article. Chapter 5 nails down the particulars of calculating statistics so that you can know what to look for and immediately tell when something's not right.

Belaboring the loss of land

In the same Sunday paper was an article about the extent of land development and speculation across the United States. Knowing how many homes are likely to be built in your neck of the woods is an important issue to get a handle on. Statistics are given regarding the number of acres of farmland being lost to development each year. To further illustrate how much land is being lost, the area is also listed in terms of football fields. In this particular example, experts said that the mid-Ohio area is losing 150,000 acres per year, which is 234 square miles, or 115,385 football fields (including end zones). How do people come up with these numbers, and how accurate are they? And does it help to visualize land loss in terms of the corresponding number of football fields? I discuss the accuracy of data collected in more detail in Chapter 16.

Scrutinizing schools

The next topic in the paper was school proficiency — specifically, whether extra school sessions help students perform better. The article states that 81.3% of students in this particular district who attended extra sessions passed the writing proficiency test, whereas only 71.7% of those who didn't participate in the extra school sessions passed it. But is this enough of a difference to account for the $386,000 price tag per year? And what's happening in these sessions to cause an improvement? Are students in these sessions spending more time just preparing for those exams rather than learning more about writing in general? And here's the big question: Were the participants in the extra sessions student volunteers who may be more motivated than the average student to try to improve their test scores? The article doesn't say.

Studying surveys of all shapes and sizes

Surveys and polls are among the most visible mechanisms used by today's media to grab your attention. It seems that everyone wants to do a survey, including market managers, insurance companies, TV stations, community groups, and even students in high school classes. Here are just a few examples of survey results that are part of today's news:

With the aging of the American workforce, companies are planning for their future leadership. (How do they know that the American workforce is aging, and if it is, by how much is it aging?) A recent survey shows that nearly 67% of human-resources managers polled said that planning for succession had become more important in the past five years than it had been in the past. The survey also says that 88% of the 210 respondents said they usually or often fill senior positions with internal candidates. But how many managers did not respond, and is 210 respondents really enough people to warrant a story on the front page of the business section? Believe it or not, when you start looking for them, you'll find numerous examples in the news of surveys based on far fewer participants than 210. (To be fair, however, 210 can

actually be a good number of subjects in some situations. The issues of what sample size is large enough and what percentage of respondents is big enough are addressed in full detail in Chapter 16.)

Some surveys are based on current interests and trends. For example, a recent Harris-Interactive survey found that nearly half (47%) of U.S. teens say their social lives would end or be worsened without their cellphones, and 57% go as far as to say that their cellphones are the key to their social life. The study also found that 42% of teens say that they can text while blindfolded (how do you really test this?). Keep in perspective, though, that the study did not tell you what percentage of teens actually have cellphones or what demographic characteristics those teens have compared to teens who do *not* have cellphones. And remember that data collected on topics like this aren't always accurate, because the individuals who are surveyed may tend to give biased answers (who wouldn't want to say they can text blindfolded?). For more information on how to interpret and evaluate the results of surveys, see Chapter 16.

Studies like this appear all the time, and the only way to know what to believe is to understand what questions to ask and to be able to critique the quality of the study. That's all part of statistics! The good news is, with a few clarifying questions, you can quickly critique statistical studies and their results. Chapter 17 helps you do just that.

Studying sports

The sports section is probably the most numerically jampacked section of the newspaper. Beginning with game scores, the win/loss percentages for each team, and the relative standing for each team, the specialized statistics reported in the sports world are so deep they require wading boots to get through. For example, basketball statistics are broken down by team, by quarter, and by player. For each player, you get minutes played, field goals, free throws, rebounds, assists, personal fouls, turnovers, blocks, steals, and total points.

Who needs to know this stuff, besides the players' mothers? Apparently many fans do. Statistics are something that sports fans can never get enough of and players often can't stand to hear about. Stats are the substance of water-cooler debates and the fuel for armchair quarterbacks around the world.

Fantasy sports have also made a huge impact on the sports money-making machine. Fantasy sports are games where participants act as owners to build their own teams from existing players in a professional league. The fantasy team owners then compete against each other. What is the competition based on? Statistical performance of the players and teams involved, as measured by rules set up by a "league commissioner" and an established point system. According to the Fantasy Sports Trade Association, the number of people age 12 and up who are involved in fantasy sports is more than 30 million, and the amount of money spent is $3–4 billion per year. (And even here you can ask how the numbers were calculated — the questions never end, do they?)

Banking on business news

The business section of the newspaper provides statistics about the stock market. In one week the market went down 455 points; is that decrease a lot or a little? You need to calculate a percentage to really get a handle on that.

The business section of my paper contained reports on the highest yields nationwide on every kind of certificate of deposit (CD) imaginable. (By the way, how do they know those yields are the highest?) I also found reports about rates on 30-year fixed loans, 15-year fixed loans, 1-year adjustable rate loans, new car loans, used car loans, home equity loans, and loans from your grandmother (well actually no, but if grandma read these statistics, she might increase her cushy rates).

Finally, I saw numerous ads for those beloved credit cards — ads listing the interest rates, the annual fees, and the number of days in the billing cycle. How do you compare all the information about investments, loans, and credit cards in order to make a good decision? What statistics are most important? The real question is: Are the numbers reported in the paper giving the whole story, or do you need to do more detective work to get at the truth? Chapters 16 and 17 help you start tearing apart these numbers and making decisions about them.

Touring the travel news

You can't even escape the barrage of numbers by heading to the travel section. For example, there I found that the most frequently asked question coming in to the Transportation Security Administration's response center (which receives about 2,000 telephone calls, 2,500 e-mail messages, and 200 letters per week on average — would you want to be the one counting all of those?) is, "Can I carry this on a plane?" *This* can refer to anything from an animal to a wedding dress to a giant tin of popcorn. (I wouldn't recommend the tin of popcorn. You have to put it in the overhead compartment horizontally, and because things shift during flight, the cover will likely open; and when you go to claim your tin at the end of the flight, you and your seatmates will be showered. Yes, I saw it happen once.)

The number of reported responses in this case leads to an interesting statistical question: How many operators are needed at various times of the day to field those calls, e-mails, and letters coming in? Estimating the number of anticipated calls is your first step, and being wrong can cost you money (if you overestimate it) or a lot of bad PR (if you underestimate it). These kinds of statistical challenges are tackled in Chapter 13.

Surveying sexual stats

In today's age of info-overkill, it's very easy to find out what the latest buzz is, including the latest research on people's sex lives. An article in my paper reported that married people have 6.9 more sexual encounters per year than people who have never been married. That's nice to know, I guess, but how did someone come up with this number? The article I'm looking at doesn't say (maybe some statistics are better left unsaid?).

If someone conducted a survey by calling people on the phone asking for a few minutes of their time to discuss their sex lives, who will be the most likely to want to talk about it? And what are they going to say in response to the question, "How many times a week do you have sex?" Are they going to report the honest truth, tell you to mind your own business, or exaggerate a little? Self-reported surveys can be a real source of bias and can lead to misleading statistics. But how would you recommend people go about finding out more about this very personal subject? Sometimes, research is more difficult than it seems. (Chapter 16 discusses biases that come up when collecting certain types of survey data.)

Breaking down weather reports

Weather reports provide another mass of statistics, with forecasts of the next day's high and low temperatures (how do they decide it'll be 16 degrees and not 15 degrees?) along with reports of the day's UV factor, pollen count, pollution standard index, and water quality and quantity. (How do they get these numbers — by taking samples? How many samples do they take, and where do they take them?) You can find out what the weather is right now anywhere in the world. You can get a forecast looking ahead three days, a week, a month, or even a year! Meteorologists collect and record tons and tons of data on the weather each day. Not only do these numbers help you decide whether to take your umbrella to work, but they also help weather researchers to better predict longer term forecasts and even global climate changes over time.

Even with all the information and technologies available to weather researchers, how accurate are weather reports these days? Given the number of times you get rained on when you were told it was going to be sunny, it seems they still have work to do on those forecasts. What the abundance of data really shows though, is that the number of variables affecting weather is almost overwhelming, not just to you, but for meteorologists, too.

Statistical computer models play an important role in making predictions about major weather-related events, such as hurricanes, earthquakes, and volcano eruptions. Scientists still have some work to do before they can predict tornados before they begin to form or tell you exactly where and when a hurricane is going to hit land, but that's certainly their goal, and they continue to get better at it. For more on modeling and statistics, see Chapter 18.

Musing about movies

Moving on to the arts section, I saw several ads for current movies. Each movie ad contains quotes from certain movie critics: "Two thumbs up!" "The supreme adventure of our time," "Absolutely hilarious," or "One of the top ten films of the year!" Do you pay attention to the critics? How do you determine which movies to go to? Experts say that although the popularity of a movie may be affected by the critics' comments (good or bad) in the beginning of a film's run, word of mouth is the most important determinant of how well a film does in the long run.

Studies also show that the more dramatic a movie is, the more popcorn is sold. Yes, the entertainment business even keeps tabs on how much crunching you do at the movies. How do they collect all this information, and how does it impact the types of movies that are made? This, too, is part of statistics: designing and carrying out studies to help pinpoint an audience and find out what they like, and then using the information to help guide the making of the product. So the next time someone with a clipboard asks if you have a minute, you may want to stand up and be counted.

Highlighting horoscopes

Those horoscopes: You read them, but do you believe them? Should you? Can people predict what will happen more often than just by chance? Statisticians have a way of finding out, by using something they call a *hypothesis test* (see Chapter 14). So far they haven't found anyone who can read minds, but people still keep trying!

Using Statistics at Work

Now put down the Sunday newspaper and move on to the daily grind of the workplace. If you're working for an accounting firm, of course numbers are part of your daily life. But what about people like nurses, portrait studio photographers, store managers, newspaper reporters, office staff, or construction workers? Do numbers play a role in those jobs? You bet. This section gives you a few examples of how statistics creep into *every* workplace.

You don't have to go far to see how statistics weaves its way in and out of your life and work. The secret is being able to determine what it all means and what you can believe, and to be able to make sound decisions based on the real story behind numbers so you can handle and become used to the statistics of everyday life.

Delivering babies — and information

Sue works as a nurse during the night shift in the labor and delivery unit at a university hospital. She takes care of several patients in a given evening, and she does her best to accommodate everyone. Her nursing manager has told her that each time she comes on shift she should identify herself to the patient, write her name on the whiteboard in the patient's room, and ask whether the patient has any questions. Why? Because a few days after each mother leaves with her baby, the hospital gives her a phone call asking about the quality of care, what was missed, what it could do to improve its service and quality of care, and what the staff could do to ensure that the hospital is chosen over other hospitals in town. For example, surveys show that patients who know the names of their nurses feel more comfortable, ask more questions, and have a more positive experience in the hospital than those who don't know the names of their nurses. Sue's salary raises depend on her ability to follow through with the needs of new mothers. No doubt the hospital has also done a lot of research to determine the factors involved in quality of patient care well beyond nurse-patient interactions. (See Chapter 17 for in-depth info concerning medical studies.)

Posing for pictures

Carol recently started working as a photographer for a department store portrait studio; one of her strengths is working with babies. Based on the number of photos purchased by customers over the years, this store has found that people buy more posed pictures than natural-looking ones. As a result, store managers encourage their photographers to take posed shots.

A mother comes in with her baby and has a special request: "Could you please not pose my baby too deliberately? I just like his pictures to look natural." If Carol says, "Can't do that, sorry. My raises are based on my ability to pose a child well," you can bet that the mother is going to fill out that survey on quality service after this session — and not just to get $2.00 off her next sitting (if she ever comes back). Instead, Carol should show her boss the information in Chapter 16 about collecting data on customer satisfaction.

Poking through pizza data

Terry is a store manager at a local pizzeria that sells pizza by the slice. He is in charge of determining how many workers to have on staff at a given time, how many pizzas to make ahead of time to accommodate the demand, and how much cheese to order and grate, all with minimal waste of wages and ingredients. Friday night at midnight, the place is dead. Terry has five workers left and has five large pans of pizza he could throw in the oven, making about 40 slices of pizza each. Should he send two of his workers home? Should he put more pizza in the oven or hold off?

The store owner has been tracking the demand for weeks now, so Terry knows that every Friday night things slow down between 10 and 12 p.m., but then the bar crowd starts pouring in around midnight and doesn't let up until the doors close at 2:30 a.m. So Terry keeps the workers on, puts in the pizzas in 30-minute intervals from midnight on, and is rewarded with a profitable night, with satisfied customers and with a happy boss. For more information on how to make good estimates using statistics, see Chapter 13.

Statistics in the office

D.J. is an administrative assistant for a computer company. How can statistics creep into her office workplace? Easy. Every office is filled with people who want to know answers to questions, and they want someone to "Crunch the numbers," to "Tell me what this means," to "Find out if anyone has any hard data on this," or to simply say, "Does this number make any sense?" They need to know everything from customer satisfaction figures to changes in inventory during the year; from the percentage of time employees spend on e-mail to the cost of supplies for the last three years. Every workplace is filled with statistics, and D.J.'s marketability and value as an employee could go up if she's the one the head honchos turn to for help. Every office needs a resident statistician — why not let it be you?

Chapter 3

Taking Control: So Many Numbers, So Little Time

*T*he sheer amount of statistics in daily life can leave you feeling overwhelmed and confused. This chapter gives you a tool to help you deal with statistics: skepticism! Not radical skepticism like "I can't believe anything anymore," but healthy skepticism like "Hmm, I wonder where that number came from?" and "I need to find out more information before I believe these results." To develop healthy skepticism, you need to understand how the chain of statistical information works.

Statistics end up on your TV and in your newspaper as a result of a process. First, the researchers who study an issue generate results; this group is composed of pollsters, doctors, marketing researchers, government researchers, and other scientists. They are considered the *original sources* of the statistical information.

After they get their results, these researchers naturally want to tell people about it, so they typically either put out a press release or publish a journal article. Enter the journalists or reporters, who are considered the *media sources* of the information. Journalists hunt for interesting press releases and sort through journals, basically searching for the next headline. When reporters complete their stories, statistics are immediately sent out to the public through all forms of media. Now the information is ready to be taken in by the third group — the *consumers* of the information (you). You and other consumers of information are faced with the task of listening to and reading the information, sorting through it, and making decisions about it.

At any stage in the process of doing research, communicating results, or consuming information, errors can take place, either unintentionally or by design. The tools and strategies you find in this chapter give you the skills to be a good detective.

Detecting Errors, Exaggerations, and Just Plain Lies

Statistics can go wrong for many different reasons. First, a simple, honest error can occur. This can happen to anyone, right? Other times, the error is something other than a simple, honest mistake. In the heat of the moment, because someone feels strongly about a cause and because the numbers don't quite bear out the point that the researcher wants to make, statistics get tweaked, or, more commonly, exaggerated, either in their values or how they're represented and discussed.

Another type of error is an *error of omission* — information that is missing that would have made a big difference in terms of getting a handle on the real story behind the numbers. That omission makes the issue of correctness difficult to address, because you're lacking information to go on.

You may even encounter situations in which the numbers have been completely fabricated and can't be repeated by anyone because they never happened. This section gives you tips to help you spot errors, exaggerations, and lies, along with some examples of each type of error that you, as an information consumer, may encounter.

Checking the math

The first thing you want to do when you come upon a statistic or the result of a statistical study is to ask, "Is this number correct?" Don't assume it is! You'd probably be surprised at the number of simple arithmetic errors that occur when statistics are collected, summarized, reported, or interpreted.

To spot arithmetic errors or omissions in statistics:

- **Check to be sure everything adds up.** In other words, do the percents in the pie chart add up to 100 (or close enough due to rounding)? Do the number of people in each category add up to the total number surveyed?

- **Double-check even the most basic calculations.**

- **Always look for a total so you can put the results into proper perspective.** Ignore results based on tiny sample sizes.

- **Examine whether the projections are reasonable.** For example, if three deaths due to a certain condition are said to happen per minute, that adds up to over 1.5 million such deaths in a year. Depending on what condition is being reported, this number may be unreasonable.

Uncovering misleading statistics

By far, the most common abuses of statistics are subtle, yet effective, exaggerations of the truth. Even when the math checks out, the underlying statistics themselves can be misleading if they exaggerate the facts. Misleading statistics are harder to pinpoint than simple math errors, but they can have a huge impact on society, and, unfortunately, they occur all the time.

Breaking down statistical debates

Crime statistics are a great example of how statistics are used to show two sides of a story, only one of which is really correct. Crime is often discussed in political debates, with one candidate (usually the incumbent) arguing that crime has gone down during her tenure, and the challenger often arguing that crime has gone up (giving the challenger something to criticize the incumbent for). How can two candidates make such different conclusions based on the same data set? Turns out, depending on the way you measure crime, getting either result can be possible.

Table 3-1 shows the population of the United States for 1998 to 2008, along with the number of reported crimes and the crime *rates* (crimes per 100,000 people), calculated by taking the number of crimes divided by the population size and multiplying by 100,000.

Table 3-1	Number of Crimes, Estimated Population Size, and Crime Rates in the U.S.		
Year	*No. of Crimes*	*Population Size*	*Crime Rate per 100,000 People*
1998	12,475,634	270,296,000	4,615.5
1999	11,634,378	272,690,813	4,266.5
2000	11,608,072	281,421,906	4,124.8
2001	11,876,669	285,317,559	4,162.6
2002	11,878,954	287,973,924	4,125.0
2003	11,826,538	290,690,788	4,068.4
2004	11,679,474	293,656,842	3,977.3
2005	11,565,499	296,507,061	3,900.6
2006	11,401,511	299,398,484	3,808.1
2007	11,251,828	301,621,157	3,730.5
2008	11,149,927	304,059,784	3,667.0

Source: U.S. Crime Victimization Survey

Now compare the number of crimes and the crime rates for 2001 and 2002 in Table 3-1. In column 2, you see that the *number of crimes* increased by 2,285 from 2001 to 2002 (11,878,954 – 11,876,669). This represents an increase of 0.019% (dividing the difference, 2,285, by the number of crimes in 2001, 11,876,669). Note the population size (column 3) also increased from 2001 to 2002, by 2,656,365 people (287,973,924 – 285,317,559), or 0.931% (dividing this difference by the population size in 2001). However, in column 4, you see the crime *rate* decreased from 2001 to 2002 from 4,162.6 (per 100,000 people) in 2001 to 4,125.0 (per 100,000) in 2002. How did the crime rate decrease? Although the number of crimes and the number of people both went up, the number of crimes increased at a slower rate than the increase in population size did (0.019% compare to 0.931%).

So how should the crime trend be reported? Did crime actually go up or down from 2001 to 2002? Based on the crime rate — which is a more accurate gauge — you can conclude that crime decreased during that year. But be watchful of the politician who wants to show that the incumbent didn't do his job; he will be tempted to look at the number of crimes and claim that crime went up, creating an artificial controversy and resulting in confusion (not to mention skepticism) on behalf of the voters. (Aren't election years fun?)

To create an even playing field when measuring how often an event occurs, you convert each number to a percent by dividing by the total to get what statisticians call a *rate*. Rates are usually better than count data because rates allow you to make fair comparisons when the totals are different.

Untwisting tornado statistics

Which state has the most tornados? It depends on how you look at it. If you just count the number of tornados in a given year (which is how I've seen the media report it most often), the top state is Texas. But think about it. Texas is the second biggest state (after Alaska). Yes, Texas is in that part of the U.S. called "Tornado Alley," and yes, it gets a lot of tornados, but it also has a huge surface area for those tornados to land and run.

A more fair comparison, and how meteorologists look at it, is to look at the number of tornados per 10,000 square miles. Using this statistic (depending on your source), Florida comes out on top, followed by Oklahoma, Indiana, Iowa, Kansas, Delaware, Louisiana, Mississippi, and Nebraska, and finally Texas weighs in at number 10. (Although I'm sure this is one statistic they are happy to rank low on; as opposed to their AP rankings in NCAA football.)

Other tornado statistics measured and reported include the state with the highest percentage of killer tornadoes as a percentage of all tornados (Tennessee); and the total length of tornado paths per 10,000 square miles (Mississippi). Note each of these statistics is reported appropriately as a *rate* (amount per unit).

Before believing statistics indicating "the highest XXX" or "the lowest XXX," take a look at how the variable is measured to see whether it's fair and whether there are other statistics that should be examined too to get the whole picture. Also make sure the units are appropriate for making fair comparisons.

Zeroing in on what the scale tells you

Charts and graphs are useful for making a quick and clear point about your data. Unfortunately, many times the charts and graphs accompanying everyday statistics aren't done correctly and/or fairly. One of the most important elements to watch for is the way that the chart or graph is scaled. The *scale* of a graph is the quantity used to represent each tick mark on the axis of the graph. Do the tick marks increase by 1s, 10s, 20s, 100s, 1,000s, or what? The scale can make a big difference in terms of the way the graph or chart looks.

For example, the Kansas Lottery routinely shows its recent results from the Pick 3 Lottery. One of the statistics reported is the number of times each number (0 through 9) is drawn among the three winning numbers. Table 3-2 shows a chart of the number of times each number was drawn during 1,613 total Pick 3 games (4,839 single numbers drawn). It also reports the percentage of times that each number was drawn. Depending on how you choose to look at these results, you can again make the statistics appear to tell very different stories.

Table 3-2	Numbers Drawn in the Pick 3 Lottery	
Number Drawn	*No. of Times Drawn out of 4,839*	*Percentage of Times Drawn (No. of Times Drawn ÷ 4,839)*
0	485	10.0%
1	468	9.7%
2	513	10.6%
3	491	10.1%
4	484	10.0%
5	480	9.9%
6	487	10.1%
7	482	10.0%
8	475	9.8%
9	474	9.8%

The way lotteries typically display results like those in Table 3-2 is shown in Figure 3-1a. Notice that in this chart, it seems that the number 1 doesn't get drawn nearly as often (only 468 times) as number 2 does (513 times). The difference in the height of these two bars appears to be very large, exaggerating the difference in the number of times these two numbers were drawn. However, to put this in perspective, the actual difference here is 513 – 468 = 45 out of a total of 4,839 numbers drawn. In terms of percentages, the difference between the number of times the number 1 and the number 2 are drawn is 45 ÷ 4,839 = 0.009, or only nine-tenths of one percent.

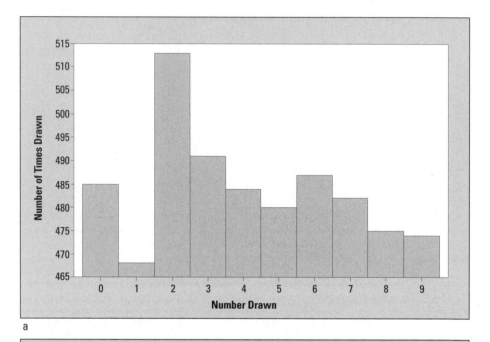

a

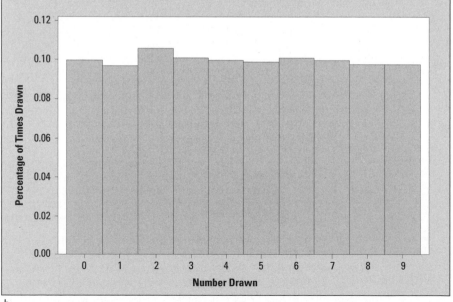

b

Figure 3-1: Bar charts showing a) number of times each number was drawn; and b) percentage of times each number was drawn.

What makes this chart exaggerate the differences? Two issues come to mind. First, notice that the vertical axis, which shows the number of times (or frequency) that each number is drawn, goes up by 5s. So a difference of 5 out of a total of 4,839 numbers drawn appears significant. Stretching the scale so that differences appear larger than they really are is a common trick used to exaggerate results. Second, the chart starts counting at 465, not at 0. Only the top part of each bar is shown, which also exaggerates the results. In comparison, Figure 3-1b graphs the *percentage* of times each number was drawn. Normally the shape of a graph wouldn't change when going from counts to percentages; however, this chart uses a more realistic scale than the one in Figure 3-1a (going by 2% increments) and starts at 0, both of which make the differences appear as they really are — not much different at all. Boring, huh?

Maybe the lottery folks thought so too. In fact, maybe they use Figure 3-1a rather than Figure 3-1b because they want you to think that some "magic" is involved in the numbers, and you can't blame them; that's their business.

Looking at the scale of a graph or chart can really help you keep the reported results in proper perspective. Stretching the scale out or starting the *y*-axis at the highest possible number makes differences appear larger; squeezing down the scale or starting the *y*-axis at a much lower value than needed makes differences appear smaller than they really are.

Checking your sources

When examining the results of any study, check the source of the information. The best results are often published in reputable journals that are well known by the experts in the field. For example, in the world of medical science, the *Journal of the American Medical Association* (JAMA), the *New England Journal of Medicine, The Lancet,* and the *British Medical Journal* are all reputable journals doctors use to publish results and read about new findings.

Consider the source and who financially supported the research. Many companies finance research and use it for advertising their products. Although that in itself isn't necessarily a bad thing, in some cases a conflict of interest on the part of researchers can lead to biased results. And if the results are very important to you, ask whether more than one study was conducted, and if so, ask to examine all the studies that were conducted, not just those whose results were published in journals or appeared in advertisements.

Counting on sample size

Sample size isn't everything, but it does count for a great deal in surveys and studies. If the study is designed and conducted correctly, and if the participants are selected randomly (that is, with no bias; see Chapter 16 for more on random samples), sample size is an important factor in determining the accuracy and repeatability of the results. (See Chapters 16 and 17 for more information on designing and carrying out studies.)

Many surveys are based on large numbers of participants, but that isn't always true for other types of research, such as carefully controlled experiments. Because of the high cost of some types of research in terms of time and money, some studies are based on a small number of participants or products. Researchers have to find the appropriate balance when determining sample size.

The most unreliable results are those based on *anecdotes,* stories that talk about a single incident in an attempt to sway opinion. Have you ever told someone not to buy a product because you had a bad experience with it? Remember that an anecdote (or story) is really a nonrandom sample whose size is only one.

Considering cause and effect

Headlines often simplify or skew the "real" information, especially when the stories involve statistics and the studies that generated the statistics.

A study conducted a few years back evaluated videotaped sessions of 1,265 patient appointments with 59 primary-care physicians and 6 surgeons in Colorado and Oregon. This study found that physicians who had not been sued for malpractice spent an average of 18 minutes with each patient, compared to 16 minutes for physicians who *had* been sued for malpractice. The study was reported by the media with the headline, "Bedside manner fends off malpractice suits." However, this study seemed to say that if you are a doctor who gets sued, all you have to do is spend more time with your patients, and you're off the hook. (Now when did bedside manner get characterized as time spent?)

Beyond that, are we supposed to believe that a doctor who has been sued needs only add a couple more minutes of time with each patient to avoid being sued in the future? Maybe what the doctor does during that time counts much more than how much time the doctor actually spends with each patient. You tackle the issues of cause-and-effect relationships between variables in Chapter 18.

Finding what you wanted to find

You may wonder how two political candidates can discuss the same topic and get two opposing conclusions, both based on "scientific surveys." Even small differences in a survey can create big differences in results. (See Chapter 16 for the full scoop on surveys.)

One common source of skewed survey results comes from question wording. Here are three different questions that are trying to get at the same issue — public opinion regarding the line-item veto option available to the president:

✔ Should the line-item veto be available to the president to eliminate waste (yes/no/no opinon)?

✔ Does the line-item veto give the president too much individual power (yes/no/no opinion)?

✔ What is your opinion on the presidential line-item veto? Choose 1–5, with 1 = strongly opposed and 5 = strongly support.

The first two questions are misleading and will lead to biased results in opposite directions. The third version will draw results that are more accurate in terms of what people really think. However, not all surveys are written with the purpose of finding the truth; many are written to support a certain viewpoint.

Research shows that even small changes in wording affect survey outcomes, leading to results that conflict when different surveys are compared. If you can tell from the wording of the question how they want you to respond to it, you know you're looking at a leading question; and leading questions lead to biased results.(See Chapter 16 for more on spotting problems with surveys.)

Looking for lies in all the right places

Every once in a while, you hear about someone who faked his data, or "fudged the numbers." Probably the most commonly committed lie involving statistics and data is when people throw out data that don't fit their hypothesis, don't fit the pattern, or appear to be outliers. In cases when someone has clearly made an error (for example, someone's age is recorded as 200), removing that erroneous data point or trying to correct the error makes sense. Eliminating data for any other reason is ethically wrong; yet it happens.

Regarding missing data from experiments, a commonly used phrase is "Among those who completed the study. . . ." What about those who didn't complete the study, especially a medical one? Did they get tired of the side effects of the experimental drug and quit? If so, the loss of this person will create results that are biased toward positive outcomes.

Before believing the results of a study, check out how many people were chosen to participate, how many finished the study, and what happened to all the participants, not just the ones who experienced a positive result.

Surveys are not immune to problems from missing data, either. For example, it's known by statisticians that the opinions of people who respond to a survey can be very different from the opinions of those who don't. In general, the lower the percentage of people who respond to a survey (the response rate), the less credible the results will be. For more about surveys and missing data, see Chapter 16.

Feeling the Impact of Misleading Statistics

You make decisions every day based on statistics and statistical studies that you've heard about or seen, many times without even realizing it. Misleading statistics affect your life in small or large ways, depending on the type of statistics that cross your path and what you choose to do with the information you're given. Here are some little everyday scenarios where statistics slip in:

- ✔ "Gee, I hope Rex doesn't chew up my rugs again while I'm at work. I heard somewhere that dogs on Prozac deal better with separation anxiety. How did they figure that out? And what would I tell my friends?"

- ✔ "I thought everyone was supposed to drink eight glasses of water a day, but now I hear that too much water could be bad for me; what should I believe?"

- ✔ "A study says people spend two hours a day at work checking and sending personal e-mails. How is that possible? No wonder my boss is paranoid."

You may run into other situations involving statistics that can have a larger impact on your life, and you need to be able to sort it all out. Here are some examples:

- ✔ A group lobbying for a new skateboard park tells you 80% of the people surveyed agree that taxes should be raised to pay for it, so you should too. Will you feel the pressure to say yes?

- ✔ The radio news at the top of the hour says cellphones cause brain tumors. Your spouse uses his cellphone all the time. Should you panic and throw away all cellphones in your house?

- ✔ You see an advertisement that tells you a certain drug will cure your particular ill. Do you run to your doctor and demand a prescription?

Although not all statistics are misleading and not everyone is out to get you, you do need to be vigilant. By sorting out the good information from the suspicious and bad information, you can steer clear of statistics that go wrong. The tools and strategies in this chapter are designed to help you to stop and say, "Wait a minute!" so you can analyze and critically think about the issues and make good decisions.

Chapter 4

Tools of the Trade

· ·

· ·

*I*n today's world, the buzzword is *data,* as in, "Do you have any data to support your claim?" "What data do you have on this?" "The data supported the original hypothesis that . . . ," "Statistical data show that . . . ," and "The data bear this out . . . ." But the field of statistics is not just about data.

Statistics is the entire process involved in gathering evidence to answer questions about the world, in cases where that evidence happens to be data.

In this chapter, you see firsthand how statistics works as a process and where the numbers play their part. You're also introduced to the most commonly used forms of statistical jargon, and you find out how these definitions and concepts all fit together as part of that process. So the next time you hear someone say, "This survey had a margin of error of plus or minus 3 percentage points," you'll have a basic idea of what that means.

Statistics: More than Just Numbers

Statisticians don't just "do statistics." Although the rest of the world views them as number crunchers, they think of themselves as the keepers of the scientific method. Of course, statisticians work with experts in other fields to satisfy their need for data, because man cannot live by statistics alone, but crunching someone's data is only a small part of a statistician's job. (In fact, if that's all we did all day, we'd quit our day jobs and moonlight as casino consultants.) In reality, statistics is involved in every aspect of the *scientific method* — formulating good questions, setting up studies, collecting good data, analyzing the data properly, and making appropriate conclusions. But aside from analyzing the data properly, what do any of these aspects have to do with statistics? In this chapter you find out.

All research starts with a question, such as:

- ✓ Is it possible to drink too much water?
- ✓ What's the cost of living in San Francisco?
- ✓ Who will win the next presidential election?
- ✓ Do herbs really help maintain good health?
- ✓ Will my favorite TV show get renewed for next year?

None of these questions asks anything directly about numbers. Yet each question requires the use of data and statistical processes to come up with the answer.

Suppose a researcher wants to determine who will win the next U.S. presidential election. To answer with confidence, the researcher has to follow several steps:

1. **Determine the population to be studied.**

   In this case, the researcher intends to study registered voters who plan to vote in the next election.

2. **Collect the data.**

   This step is a challenge, because you can't go out and ask every person in the United States whether they plan to vote, and if so, for whom they plan to vote. Beyond that, suppose someone says, "Yes, I plan to vote." Will that person *really* vote come Election Day? And will that same person tell you whom he actually plans to vote for? And what if that person changes his mind later on and votes for a different candidate?

3. **Organize, summarize, and analyze the data.**

   After the researcher has gone out and collected the data she needs, getting it organized, summarized, and analyzed helps the researcher answer her question. This step is what most people recognize as the business of statistics.

4. **Take all the data summaries, charts, graphs, and analyses and draw conclusions from them to try to answer the researcher's original question.**

   Of course, the researcher will not be able to have 100% confidence that her answer is correct, because not every person in the United States was asked. But she can get an answer that she is *nearly* 100% sure is correct. In fact, with a sample of about 2,500 people who are selected in a fair and *unbiased* way (that is, every possible sample of size 2,500 had an equal chance of being selected), the researcher can get accurate results within plus or minus 2.5% (if all the steps in the research process are done correctly).

In making conclusions, the researcher has to be aware that every study has limits and that — because the chance for error always exists — the results could be wrong. A numerical value can be reported that tells others how confident the researcher is about the results and how accurate these results are expected to be. (See Chapter 12 for more information on margin of error.)

After the research is done and the question has been answered, the results typically lead to even more questions and even more research. For example, if men appear to favor one candidate but women favor the opponent, the next questions may be: "Who goes to the polls more often on Election Day — men or women — and what factors determine whether they will vote?"

The field of statistics is really the business of using the scientific method to answer research questions about the world. Statistical methods are involved in every step of a good study, from designing the research to collecting the data, organizing and summarizing the information, doing an analysis, drawing conclusions, discussing limitations, and, finally, designing the next study in order to answer new questions that arise. Statistics is more than just numbers — it's a process.

Grabbing Some Basic Statistical Jargon

Every trade has a basic set of tools, and statistics is no different. If you think about the statistical process as a series of stages that you go through to get from question to answer, you may guess that at each stage you'll find a group of tools and a set of terms (or jargon) to go along with it. Now if the hair is beginning to stand up on the back of your neck, don't worry. No one is asking you to become a statistics expert and plunge into the heavy-duty stuff, or to turn into a statistics nerd who uses this jargon all the time. Hey, you don't even have to carry a calculator and pocket protector in your shirt pocket (because statisticians really don't do that; it's just an urban myth).

But as the world becomes more numbers-conscious, statistical terms are thrown around more in the media and in the workplace, so knowing what the language really means can give you a leg up. Also, if you're reading this book because you want to find out more about how to calculate some statistics, understanding basic jargon is your first step. So, in this section, you get a taste of statistical jargon; I send you to the appropriate chapters elsewhere in the book to get details.

Data

Data are the actual pieces of information that you collect through your study. For example, I asked five of my friends how many pets they own, and the data they gave me are the following: 0, 2, 1, 4, 18. (The fifth friend counted each of her aquarium fish as a separate pet.) Not all data are numbers; I also recorded the gender of each of my friends, giving me the following data: male, male, female, male, female.

Most data fall into one of two groups: numerical or categorical. (I present the main ideas about these variables here; see Chapter 5 for more details.)

- ✔ **Numerical data:** These data have meaning as a measurement, such as a person's height, weight, IQ, or blood pressure; or they're a count, such as the number of stock shares a person owns, how many teeth a dog has, or how many pages you can read of your favorite book before you fall asleep. (Statisticians also call numerical data *quantitative data*.)

  Numerical data can be further broken into two types: discrete and continuous.

  - *Discrete data* represent items that can be counted; they take on possible values that can be listed out. The list of possible values may be fixed (also called *finite*); or it may go from 0, 1, 2, on to infinity (making it *countably infinite*). For example, the number of heads in 100 coin flips takes on values from 0 through 100 (finite case), but the number of flips needed to get 100 heads takes on values from 100 (the fastest scenario) on up to infinity. Its possible values are listed as 100, 101, 102, 103, . . . (representing the countably infinite case).

  - *Continuous data* represent measurements; their possible values cannot be counted and can only be described using intervals on the real number line. For example, the exact amount of gas purchased at the pump for cars with 20-gallon tanks represents nearly-continuous data from 0.00 gallons to 20.00 gallons, represented by the interval [0, 20], inclusive. (Okay, you *can* count all these values, but why would you want to? In cases like these, statisticians bend the definition of continuous a wee bit.) The lifetime of a C battery can be anywhere from 0 to infinity, technically, with all possible values in between. Granted, you don't expect a battery to last more than a few hundred hours, but no one can put a cap on how long it can go (remember the Energizer Bunny?).

- ✔ **Categorical data:** Categorical data represent characteristics such as a person's gender, marital status, hometown, or the types of movies they like. Categorical data can take on numerical values (such as "1" indicating male and "2" indicating female), but those numbers don't have meaning. You couldn't add them together, for example. (Other names for categorical data are *qualitative data*, or *Yes/No data*.)

Ordinal data mixes numerical and categorical data. The data fall into categories, but the numbers placed on the categories have meaning. For example,
rating a restaurant on a scale from 0 to 4 stars gives ordinal data. Ordinal data
are often treated as categorical, where the groups are ordered when graphs
and charts are made. I don't address them separately in this book.

Data set

A *data set* is the collection of all the data taken from your sample. For example, if you measured the weights of five packages, and those weights were
12, 15, 22, 68, and 3 pounds, those five numbers (12, 15, 22, 68, 3) constitute
your data set. If you only record the general size of the package (for example,
small, medium, or large), your data set may look like this: medium, medium,
medium, large, small.

Variable

A *variable* is any characteristic or numerical value that varies from individual
to individual. A variable can represent a count (for example, the number of
pets you own); or a measurement (the time it takes you to wake up in the
morning). Or the variable can be categorical, where each individual is placed
into a group (or category) based on certain criteria (for example, political
affiliation, race, or marital status). Actual pieces of information recorded on
individuals regarding a variable are the data.

Population

For virtually any question you may want to investigate about the world, you
have to center your attention on a particular group of individuals (for example,
a group of people, cities, animals, rock specimens, exam scores, and so on).
For example:

 ✔ What do Americans think about the president's foreign policy?

 ✔ What percentage of planted crops in Wisconsin did deer destroy last year?

 ✔ What's the prognosis for breast cancer patients taking a new experimental drug?

 ✔ What percentage of all cereal boxes get filled according to specification?

In each of these examples, a question is posed. And in each case, you can identify a specific group of individuals being studied: the American people, all planted crops in Wisconsin, all breast cancer patients, and all cereal boxes that are being filled, respectively. The group of individuals you want to study in order to answer your research question is called a *population*. Populations, however, can be hard to define. In a good study, researchers define the population very clearly, whereas in a bad study, the population is poorly defined.

The question of whether babies sleep better with music is a good example of how difficult defining the population can be. Exactly how would you define a baby? Under three months old? Under a year? And do you want to study babies only in the United States, or all babies worldwide? The results may be different for older and younger babies, for American versus European versus African babies, and so on.

 Many times researchers want to study and make conclusions about a broad population, but in the end — to save time, money, or just because they don't know any better — they study only a narrowly defined population. That short-cut can lead to big trouble when conclusions are drawn. For example, suppose a college professor wants to study how TV ads persuade consumers to buy products. Her study is based on a group of her own students who partici-pated to get five points extra credit. This test group may be convenient, but her results can't be generalized to any population beyond her own students, because no other population was represented in her study.

Sample, random, or otherwise

When you sample some soup, what do you do? You stir the pot, reach in with a spoon, take out a little bit of the soup, and taste it. Then you draw a conclu-sion about the whole pot of soup, without actually having tasted all of it. If your sample is taken in a fair way (for example, you didn't just grab all the good stuff) you will get a good idea how the soup tastes without having to eat it all. Taking a sample works the same way in statistics. Researchers want to find out something about a population, but they don't have time or money to study every single individual in the population. So they select a subset of individuals from the population, study those individuals, and use that infor-mation to draw conclusions about the whole population. This subset of the population is called a *sample*.

Although the idea of a selecting a sample seems straightforward, it's anything but. The way a sample is selected from the population can mean the difference between results that are correct and fair and results that are garbage. Example: Suppose you want a sample of teenagers' opinions on whether they're spend-ing too much time on the Internet. If you send out a survey using text messag-ing, your results won't represent the opinions of *all teenagers,* which is your intended population. They will represent only those teenagers who have access to text messages. Does this sort of statistical mismatch happen often? You bet.

Some of the biggest culprits of statistical misrepresentation caused by bad sampling are surveys done on the Internet. You can find thousands of surveys on the Internet that are done by having people log on to a particular Web site and give their opinions. But even if 50,000 people in the U.S. complete a survey on the Internet, it doesn't represent the population of all Americans. It represents only those folks who have Internet access, who logged on to that particular Web site, and who were interested enough to participate in the survey (which typically means that they have strong opinions about the topic in question). The result of all these problems is *bias* — systematic favoritism of certain individuals or certain outcomes of the study.

How do you select a sample in a way that avoids bias? The key word is *random.* A *random sample* is a sample selected by equal opportunity; that is, every possible sample the same size as yours had an equal chance to be selected from the population. What *random* really means is that no group in the population is favored in or excluded from the selection process.

Non-random (in other words *bad*) *samples* are samples that were selected in such a way that some type of favoritism and/or automatic exclusion of a part of the population was involved. A classic example of a non-random sample comes from polls for which the media asks you to phone in your opinion on a certain issue ("call-in" polls). People who choose to participate in call-in polls do not represent the population at large because they had to be watching that program, and they had to feel strongly enough to call in. They technically don't represent a sample at all, in the statistical sense of the word, because no one selected them beforehand — they selected themselves to participate, creating a *volunteer* or *self-selected* sample. The results will be skewed toward people with strong opinions.

To take an authentic random sample, you need a randomizing mechanism to select the individuals. For example, the Gallup Organization starts with a computerized list of all telephone exchanges in America, along with estimates of the number of residential households that have those exchanges. The computer uses a procedure called *random digit dialing* (RDD) to randomly create phone numbers from those exchanges, and then selects samples of telephone numbers from those. So what really happens is that the computer creates a list of *all possible* household phone numbers in America and then selects a subset of numbers from that list for Gallup to call.

Another example of random sampling involves the use of random number generators. In this process, the items in the sample are chosen using a computer-generated list of random numbers, where each sample of items has the same chance of being selected. Researchers may use this type of randomization to assign patients to a treatment group versus a control group in an experiment. This process is equivalent to drawing names out of a hat or drawing numbers in a lottery.

No matter how large a sample is, if it's based on non-random methods, the results will not represent the population that the researcher wants to draw conclusions about. Don't be taken in by large samples — first check to see how they were selected. Look for the term *random sample*. If you see that term, dig further into the fine print to see how the sample was actually selected and use the preceding definition to verify that the sample was, in fact, selected randomly. A small random sample is better than a large non-random one.

Statistic

A *statistic* is a number that summarizes the data collected from a sample. People use many different statistics to summarize data. For example, data can be summarized as a percentage (60% of U.S. households sampled own more than two cars), an average (the average price of a home in this sample is . . .), a median (the median salary for the 1,000 computer scientists in this sample was . . .), or a percentile (your baby's weight is at the 90th percentile this month, based on data collected from over 10,000 babies).

The type of statistic calculated depends on the type of data. For example, percentages are used to summarize categorical data, and means are used to summarize numerical data. The price of a home is a numerical variable, so you can calculate its mean or standard deviation. However, the color of a home is a categorical variable; finding the standard deviation or median of color makes no sense. In this case, the important statistics are the percentages of homes of each color.

Not all statistics are correct or fair, of course. Just because someone gives you a statistic, nothing guarantees that the statistic is scientific or legitimate. You may have heard the saying, "Figures don't lie, but liars figure."

Parameter

Statistics are based on sample data, not on population data. If you collect data from the entire population, that process is called a *census.* If you then summarize the entire census information from one variable into a single number, that number is a *parameter,* not a statistic. Most of the time, researchers are trying to estimate the parameters using statistics. The U.S. Census Bureau wants to report the total number of people in the U.S., so it conducts a census. However, due to logistical problems in doing such an arduous task (such as being able to contact homeless folks), the census numbers can only be called *estimates* in the end, and they're adjusted upward to account for people the census missed.

Bias

Bias is a word you hear all the time, and you probably know that it means something bad. But what really constitutes bias? *Bias* is systematic favoritism that is present in the data collection process, resulting in lopsided, misleading results. Bias can occur in any of a number of ways:

- ✔ **In the way the sample is selected:** For example, if you want to estimate how much holiday shopping people in the United States plan to do this year, and you take your clipboard and head out to a shopping mall on the day after Thanksgiving to ask customers about their shopping plans, you have bias in your sampling process. Your sample tends to favor those die-hard shoppers at that particular mall who were braving the massive crowds on that day known to retailers and shoppers as "Black Friday."

- ✔ **In the way data are collected:** Poll questions are a major source of bias. Because researchers are often looking for a particular result, the questions they ask can often reflect and lead to that expected result. For example, the issue of a tax levy to help support local schools is something every voter faces at one time or another. A poll question asking, "Don't you think it would be a great investment in our future to support the local schools?" has a bit of bias. On the other hand, so does "Aren't you tired of paying money out of your pocket to educate other people's children?" Question wording can have a huge impact on results.

Other issues that result in bias with polls are timing, length, level of question difficulty, and the manner in which the individuals in the sample were contacted (phone, mail, house-to-house, and so on). See Chapter 16 for more information on designing and evaluating polls and surveys.

When examining polling results that are important to you or that you're particularly interested in, find out what questions were asked and exactly how the questions were worded before drawing your conclusions about the results.

Mean (Average)

The mean, also referred to by statisticians as the *average,* is the most common statistic used to measure the center, or middle, of a numerical data set. The *mean* is the sum of all the numbers divided by the total number of numbers. The mean of the entire population is called the *population mean,* and the mean of a sample is called the *sample mean.* (See Chapter 5 for more on the mean.)

The mean may not be a fair representation of the data, because the average is easily influenced by *outliers* (very small or large values in the data set that are not typical).

Median

The median is another way to measure the center of a numerical data set. A statistical median is much like the median of an interstate highway. On many highways, the median is the middle, and an equal number of lanes lay on either side of it. In a numerical data set, the *median* is the point at which there are an equal number of data points whose values lie above and below the median value. Thus, the median is truly the middle of the data set. See Chapter 5 for more on the median.

 The next time you hear an average reported, look to see whether the median is also reported. If not, ask for it! The average and the median are two different representations of the middle of a data set and can often give two very different stories about the data, especially when the data set contains outliers (very large or small numbers that are not typical).

Standard deviation

Have you heard anyone report that a certain result was found to be "two standard deviations above the mean"? More and more, people want to report how significant their results are, and the number of standard deviations above or below average is one way to do it. But exactly what is a standard deviation?

The *standard deviation* is a measurement statisticians use for the amount of variability (or spread) among the numbers in a data set. As the term implies, a standard deviation is a standard (or typical) amount of deviation (or distance) from the average (or mean, as statisticians like to call it). So the standard deviation, in very rough terms, is the average distance from the mean.

The formula for standard deviation (denoted by s) is as follows, where n equals the number of values in the data set, each x represents a number in the data set, and $\bar{x}$ is the average of all the data:

$$s = \sqrt{\sum \frac{(x-\bar{x})^2}{n-1}}$$

For detailed instructions on calculating the standard deviation, see Chapter 5.

The standard deviation is also used to describe where most of the data should fall, in a relative sense, compared to the average. For example, if your data have the form of a bell-shaped curve (also known as a *normal distribution*), about 95% of the data lie within two standard deviations of the mean. (This result is called the *empirical rule,* or the *68–95–99.7% rule.* See Chapter 5 for more on this.)

The standard deviation is an important statistic, but it is often absent when statistical results are reported. Without it, you're getting only part of the story about the data. Statisticians like to tell the story about the man who had one foot in a bucket of ice water and the other foot in a bucket of boiling water. He said on average he felt just great! But think about the variability in the two temperatures for each of his feet. Closer to home, the average house price, for example, tells you nothing about the range of house prices you may encounter when house-hunting. The average salary may not fully represent what's really going on in your company, if the salaries are extremely spread out.

Don't be satisfied with finding out only the average — be sure to ask for the standard deviation as well. Without a standard deviation, you have no way of knowing how spread out the values may be. (If you're talking starting salaries, for example, this could be very important!)

Percentile

You've probably heard references to percentiles before. If you've taken any kind of standardized test, you know that when your score was reported, it was presented to you with a measure of where you stood compared to the other people who took the test. This comparison measure was most likely reported to you in terms of a percentile. The *percentile* reported for a given score is the percentage of values in the data set that fall below that certain score. For example, if your score was reported to be at the 90th percentile, that means that 90% of the other people who took the test with you scored lower than you did (and 10% scored higher than you did). The median is right in the middle of a data set, so it represents the 50th percentile. For more specifics on percentiles, see Chapter 5.

Percentiles are used in a variety of ways for comparison purposes and to determine *relative standing* (that is, how an individual data value compares to the rest of the group). Babies' weights are often reported in terms of percentiles, for example. Percentiles are also used by companies to see where they stand compared to other companies in terms of sales, profits, customer satisfaction, and so on.

Standard score

The standard score is a slick way to put results in perspective without having to provide a lot of details — something that the media loves. The *standard score* represents the number of standard deviations above or below the mean (without caring what that standard deviation or mean actually are).

For example, suppose Bob took his statewide 10th-grade test recently and scored 400. What does that mean? Not much, because you can't put 400 into perspective. But knowing that Bob's standard score on the test is +2 tells you everything. It tells you that Bob's score is two standard deviations above the mean. (Bravo, Bob!) Now suppose Emily's standard score is –2. In this case, this is not good (for Emily), because it means her score is two standard deviations *below* the mean.

The process of taking a number and converting it to a standard score is called *standardizing*. For the details on calculating and interpreting standard scores when you have a normal (bell-shaped) distribution, see Chapter 9.

Distribution and normal distribution

The *distribution* of a data set (or a population) is a listing or function showing all the possible values (or intervals) of the data and how often they occur. When a distribution of categorical data is organized, you see the number or percentage of individuals in each group. When a distribution of numerical data is organized, they're often ordered from smallest to largest, broken into reasonably sized groups (if appropriate), and then put into graphs and charts to examine the shape, center, and amount of variability in the data.

The world of statistics includes dozens of different distributions for categorical and numerical data; the most common ones have their own names. One of the most well-known distributions is called the *normal distribution,* also known as the *bell-shaped curve.* The normal distribution is based on numerical data that is continuous; its possible values lie on the entire real number line. Its overall shape, when the data are organized in graph form, is a symmetric bell-shape. In other words, most (around 68%) of the data are centered around the mean (giving you the middle part of the bell), and as you move farther out on either side of the mean, you find fewer and fewer values (representing the downward sloping sides on either side of the bell).

The mean (and hence the median) is directly in the center of the normal distribution due to symmetry, and the standard deviation is measured by the distance from the mean to the *inflection point* (where the curvature of the bell changes from concave up to concave down). Figure 4-1 shows a graph of a normal distribution with mean 0 and standard deviation 1 (this distribution has a special name, the *standard normal distribution* or *Z-distribution*).The shape of the curve resembles the outline of a bell.

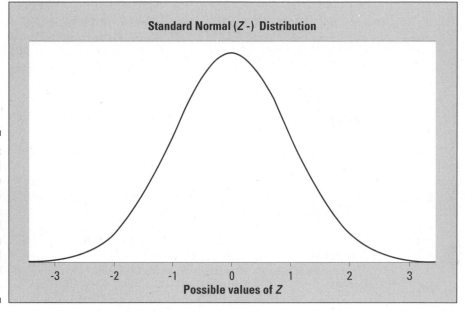

Standard Normal (*Z* -) Distribution

Possible values of *Z*

Figure 4-1:
A standard
normal (*Z*-)
distribu-
tion has a
bell-shaped
curve with
mean 0 and
standard
deviation 1.

Because every distinct population of data has a different mean and standard deviation, an infinite number of different normal distributions exist, each with its own mean and its own standard deviation to characterize it. See Chapter 9 for plenty more on the normal and standard normal distributions.

Central Limit Theorem

The normal distribution is also used to help measure the accuracy of many statistics, including the mean, using an important result in statistics called the *Central Limit Theorem.* This theorem gives you the ability to measure how much your sample mean will vary, without having to take any other sample means to compare it with (thankfully!). By taking this variability into account, you can now use your data to answer questions about the population, such as "What's the mean household income for the whole U.S.?"; or "This report said 75% of all gift cards go unused; is that really true?" (These two particular analyses made possible by the Central Limit Theorem are called *confidence intervals* and *hypothesis tests,* respectively, and are described in Chapters 13 and 14, respectively.)

The Central Limit Theorem (*CLT* for short) basically says that for non-normal data, your sample mean has an approximate normal distribution, no matter what the distribution of the original data looks like (as long as your sample size was large enough). And it doesn't just apply to the sample mean; the CLT is also true for other sample statistics, such as the sample proportion (see Chapters 13 and 14). Because statisticians know so much about the normal distribution (see the preceding section), these analyses are much easier. See Chapter 11 for more on the Central Limit Theorem, known by statisticians as the "Crown jewel in the field of all statistics." (Should you even bother to tell them to get a life?)

z-values

If a data set has a normal distribution, and you standardize all the data to obtain standard scores, those standard scores are called *z*-values. All *z*-values have what is known as a standard normal distribution (or *Z*-distribution). The *standard normal distribution* is a special normal distribution with a mean equal to 0 and a standard deviation equal to 1.

The standard normal distribution is useful for examining the data and determining statistics like percentiles, or the percentage of the data falling between two values. So if researchers determine that the data have a normal distribution, they usually first standardize the data (by converting each data point into a *z*-value) and then use the standard normal distribution to explore and discuss the data in more detail. See Chapter 9 for more details on *z*-values.

Experiments

An *experiment* is a study that imposes a treatment (or control) to the subjects (participants), controls their environment (for example, restricting their diets, giving them certain dosage levels of a drug or placebo, or asking them to stay awake for a prescribed period of time), and records the responses. The purpose of most experiments is to pinpoint a cause-and-effect relationship between two factors (such as alcohol consumption and impaired vision; or dosage level of a drug and intensity of side effects). Here are some typical questions that experiments try to answer:

- ✔ Does taking zinc help reduce the duration of a cold? Some studies show that it does.
- ✔ Does the shape and position of your pillow affect how well you sleep at night? The Emory Spine Center in Atlanta says yes.
- ✔ Does shoe heel height affect foot comfort? A study done at UCLA says up to one-inch heels are better than flat soles.

In this section, I discuss some additional definitions of words that you may hear when someone is talking about experiments. Chapter 17 is entirely dedicated to the subject. For now, just concentrate on basic experiment lingo.

Treatment group versus control group

Most experiments try to determine whether some type of experimental treatment (or important factor) has a significant effect on an outcome. For example, does zinc help to reduce the length of a cold? Subjects who are chosen to participate in the experiment are typically divided into two groups: a treatment group and a control group. (More than one treatment group is possible.)

> ✔ The *treatment group* consists of participants who receive the experimental treatment whose effect is being studied (in this case, zinc tablets).
>
> ✔ The *control group* consists of participants who do not receive the experimental treatment being studied. Instead, they get a placebo (a fake treatment; for example, a sugar pill); a standard, nonexperimental treatment (such as vitamin C, in the zinc study); or no treatment at all, depending on the situation.

In the end, the responses of those in the treatment group are compared with the responses from the control group to look for differences that are statistically significant (unlikely to have occurred just by chance).

Placebo

A *placebo* is a fake treatment, such as a sugar pill. Placebos are given to the control group to account for a psychological phenomenon called the *placebo effect,* in which patients receiving a fake treatment still report having a response, as if it were the real treatment. For example, after taking a sugar pill a patient experiencing the placebo effect might say, "Yes, I feel better already," or "Wow, I *am* starting to feel a bit dizzy." By measuring the placebo effect in the control group, you can tease out what portion of the reports from the treatment group were real and what portion were likely due to the placebo effect. (Experimenters assume that the placebo effect affects both the treatment and control groups.)

Blind and double-blind

A *blind experiment* is one in which the subjects who are participating in the study are not aware of whether they're in the treatment group or the control group. In the zinc example, the vitamin C tablets and the zinc tablets would be made to look exactly alike and patients would not be told which type of pill they were taking. A blind experiment attempts to control for bias on the part of the participants.

A *double-blind experiment* controls for potential bias on the part of both the patients *and* the researchers. Neither the patients nor the researchers collecting the data know which subjects received the treatment and which didn't. So who does know what's going on as far as who gets what treatment? Typically a third party (someone not otherwise involved in the experiment) puts together the pieces independently. A double-blind study is best, because even though researchers may claim to be unbiased, they often have a special interest in the results — otherwise they wouldn't be doing the study!

Surveys (Polls)

A *survey* (more commonly known as a *poll*) is a questionnaire; it's most often used to gather people's opinions along with some relevant demographic information. Because so many policymakers, marketers, and others want to "get at the pulse of the American public" and find out what the average American is thinking and feeling, many people now feel that they cannot escape the barrage of requests to take part in surveys and polls. In fact, you've probably received many requests to participate in surveys, and you may even have become numb to them, simply throwing away surveys received in the mail or saying "no" when asked to participate in a telephone survey.

If done properly, a good survey can really be informative. People use surveys to find out what TV programs Americans (and others) like, how consumers feel about Internet shopping, and whether the United States should allow someone under 35 to become president. Surveys are used by companies to assess the level of satisfaction their customers feel, to find out what products their customers want, and to determine who is buying their products. TV stations use surveys to get instant reactions to news stories and events, and movie producers use them to determine how to end their movies.

However, if I had to choose one word to assess the general state of surveys in the media today, I'd say it's *quantity* rather than *quality*. In other words, you'll find no shortage of bad surveys. But in this book you find no shortage of good tips and information for analyzing, critiquing, and understanding survey results, and for designing your own surveys to do the job right. (To take off with surveys, head to Chapter 16.)

Margin of error

You've probably heard or seen results like this: "This survey had a margin of error of plus or minus 3 percentage points." What does this mean? Most surveys (except a census) are based on information collected from a sample of individuals, not the entire population. A certain amount of error is bound to occur — not in the sense of calculation error (although there may be some of that, too) but in the sense of *sampling error,* which is the error that occurs simply because the researchers aren't asking everyone. The *margin of error* is supposed to measure the maximum amount by which the sample results are expected to differ from those of the actual population. Because the results of most survey questions can be reported in terms of percentages, the margin of error most often appears as a percentage, as well.

How do you interpret a margin of error? Suppose you know that 51% of people sampled say that they plan to vote for Ms. Calculation in the upcoming election. Now, projecting these results to the whole voting population, you would have to add and subtract the margin of error and give a range of possible results in order to have sufficient confidence that you're bridging the gap between your sample and the population. Supposing a margin of error of plus or minus 3 percentage points, you would be pretty confident that between 48% (51% – 3%) and 54% (51% + 3%) of the population will vote for Ms. Calculation in the election, based on the sample results. In this case, Ms. Calculation may get slightly more or slightly less than the majority of votes and could either win or lose the election. This has become a familiar situation in recent years when the media want to report results on Election Night, but based on early exit polling results, the election is "too close to call." For more on the margin of error, see Chapter 12.

The margin of error measures accuracy; it does not measure the amount of bias that may be present (find a discussion of bias earlier in this chapter). Results that look numerically scientific and precise don't mean anything if they were collected in a biased way.

Confidence interval

One of the biggest uses of statistics is to estimate a population parameter using a sample statistic. In other words, use a number that summarizes a sample to help you guesstimate the corresponding number that summarizes the whole population (the definitions of parameter and statistic appear earlier in this chapter). You're looking for a population parameter in each of the following questions:

- ✔ What's the average household income in America? (Population = all households in America; parameter = average household income.)

- ✔ What percentage of all Americans watched the Academy Awards this year? (Population = all Americans; parameter = percentage who watched the Academy Awards this year.)

- ✔ What's the average life expectancy of a baby born today? (Population = all babies born today; parameter = average life expectancy.)

- ✔ How effective is this new drug on adults with Alzheimer's? (Population = all people who have Alzheimer's; parameter = percentage of these people who see improvement when taking this drug.)

It's not possible to find these parameters exactly; they each require an estimate based on a sample. You start by taking a random sample from a population (say a sample of 1,000 households in America) and then finding the corresponding statistic from that sample (the sample's mean household income). Because you know that sample results vary from sample to sample, you need to add a "plus or minus something" to your sample results if you want to draw conclusions about the whole population (all households in

America). This "plus or minus" that you add to your sample statistic in order to estimate a parameter is the margin of error.

When you take a sample statistic (such as the sample mean or sample percentage) and add/subtract a margin of error, you come up with what statisticians call a *confidence interval*. A confidence interval represents a range of likely values for the population parameter, based on your sample statistic. For example, suppose the average time it takes you to drive to work each day is 35 minutes, with a margin of error of plus or minus 5 minutes. You estimate that the average time to work would be anywhere from 30 to 40 minutes. This estimate is a confidence interval.

Some confidence intervals are wider than others (and wide isn't good, because it equals less accuracy). Several factors influence the width of a confidence interval, such as sample size, the amount of variability in the population being studied, and how confident you want to be in your results. (Most researchers are happy with a 95% level of confidence in their results.) For more on factors that influence confidence intervals, as well as instructions for calculating and interpreting confidence intervals, see Chapter 13.

Hypothesis testing

Hypothesis test is a term you probably haven't run across in your everyday dealings with numbers and statistics. But I guarantee that hypothesis tests have been a big part of your life and your workplace, simply because of the major role they play in industry, medicine, agriculture, government, and a host of other areas. Any time you hear someone talking about their study showing a "statistically significant result," you're encountering a hypothesis test. (A statistically significant result is one that is unlikely to have occurred by chance. See Chapter 14 for the full scoop.)

Basically, a *hypothesis test* is a statistical procedure in which data are collected from a sample and measured against a claim about a population parameter. For example, if a pizza delivery chain claims to deliver all pizzas within 30 minutes of placing the order, on average, you could test whether this claim is true by collecting a random sample of delivery times over a certain period and looking at the average delivery time for that sample. To make your decision, you must also take into account the amount by which your sample results can change from sample to sample (which is related to the margin of error).

Because your decision is based on a sample and not the entire population, a hypothesis test can sometimes lead you to the wrong conclusion. However, statistics are all you have, and if done properly, they can give you a good chance of being correct. For more on the basics of hypothesis testing, see Chapter 14.

A variety of hypothesis tests are done in scientific research, including *t*-tests (comparing two population means), paired *t*-tests (looking at before/after data), and tests of claims made about proportions or means for one or more populations. For specifics on these hypothesis tests, see Chapter 15.

p-values

Hypothesis tests are used to test the validity of a claim that is made about a population. This claim that's on trial, in essence, is called the *null hypothesis.* The *alternative hypothesis* is the one you would believe if the null hypothesis is concluded to be untrue. The evidence in the trial is your data and the statistics that go along with it. All hypothesis tests ultimately use a *p*-value to weigh the strength of the evidence (what the data are telling you about the population). The *p*-value is a number between 0 and 1 and interpreted in the following way:

- ✔ A small *p*-value (typically ≤ 0.05) indicates strong evidence against the null hypothesis, so you reject it.

- ✔ A large *p*-value (> 0.05) indicates weak evidence against the null hypothesis, so you fail to reject it.

- ✔ *p*-values very close to the cutoff (0.05) are considered to be marginal (could go either way). Always report the *p*-value so your readers can draw their own conclusions.

For example, suppose a pizza place claims their delivery times are 30 minutes or less on average but you think it's more than that. You conduct a hypothesis test because you believe the null hypothesis, H_o, that the mean delivery time is 30 minutes max, is incorrect. Your alternative hypothesis (H_a) is that the mean time is greater than 30 minutes. You randomly sample some delivery times and run the data through the hypothesis test, and your *p*-value turns out to be 0.001, which is much less than 0.05. You conclude that the pizza place is wrong; their delivery times are in fact more than 30 minutes on average, and you want to know what they're gonna do about it! (Of course you could be wrong by having sampled an unusually high number of late pizzas just by chance; but whose side am I on?) For more on *p*-values, head to Chapter 14.

Statistical significance

Whenever data are collected to perform a hypothesis test, the researcher is typically looking for something out of the ordinary. (Unfortunately, research that simply confirms something that was already well known doesn't make headlines.) Statisticians measure the amount by which a result is out of the ordinary using hypothesis tests (see Chapter 14). They define a *statistically significant* result as a result with a very small probability of happening just by chance, and provide a number called a *p*-value to reflect that probability (see the previous section on *p*-values).

For example, if a drug is found to be more effective at treating breast cancer than the current treatment is, researchers say that the new drug shows a statistically significant improvement in the survival rate of patients with breast cancer. That means that based on their data, the difference in the overall results from patients on the new drug compared to those using the old treatment is so big that it would be hard to say it was just a coincidence. However, proceed with caution: You can't say that these results necessarily apply to each individual or to each individual in the same way. For full details on statistical significance, see Chapter 14.

When you hear that a study's results are statistically significant, don't automatically assume that the study's results are important. *Statistically significant* means the results were unusual, but unusual doesn't always mean important. For example, would you be excited to learn that cats move their tails more often when lying in the sun than when lying in the shade, and that those results are statistically significant? This result may not even be important to the cat, much less anyone else!

Sometimes statisticians make the wrong conclusion about the null hypothesis because a sample doesn't represent the population (just by chance). For example, a positive effect that's experienced by a sample of people who took the new treatment may have just been a fluke; or in the example in the preceding section, the pizza company really was delivering those pizzas on time and you just got an unlucky sample of slow ones. However, the beauty of research is that as soon as someone gives a press release saying that she found something significant, the rush is on to try to replicate the results, and if the results can't be replicated, this probably means that the original results were wrong for some reason (including being wrong just by chance). Unfortunately, a press release announcing a "major breakthrough" tends to get a lot of play in the media, but follow-up studies refuting those results often don't show up on the front page.

One statistically significant result shouldn't lead to quick decisions on anyone's part. In science, what most often counts is not a single remarkable study, but a body of evidence that is built up over time, along with a variety of well-designed follow-up studies. Take any major breakthroughs you hear about with a grain of salt and wait until the follow-up work has been done before using the information from a single study to make important decisions in your life. The results may not be replicable, and even if they are, you can't know if they necessarily apply to each individual.

Correlation versus causation

Of all of the misunderstood statistical issues, the one that's perhaps the most problematic is the misuse of the concepts of correlation and causation.

Correlation, as a statistical term, is the extent to which two numerical variables have a linear relationship (that is, a relationship that increases or decreases at a constant rate). Following are three examples of correlated variables:

- The number of times a cricket chirps per second is strongly related to temperature; when it's cold outside, they chirp less frequently, and as the temperature warms up, they chirp at a steadily increasing rate. In statistical terms, you say number of cricket chirps and temperature have a strong positive correlation.

- The number of crimes (per capita) has often been found to be related to the number of police officers in a given area. When more police officers patrol the area, crime tends to be lower, and when fewer police officers are present in the same area, crime tends to be higher. In statistical terms we say the number of police officers and the number of crimes have a strong negative correlation.

- The consumption of ice cream (pints per person) and the number of murders in New York are positively correlated. That is, as the amount of ice cream sold per person increases, the number of murders increases. Strange but true!

But correlation as a statistic isn't able to explain *why* or *how* the relationship between two variables, x and $y,$ exists; only that it does exist.

Causation goes a step further than correlation, stating that a change in the value of the x variable *will cause* a change in the value of the y variable. Too many times in research, in the media, or in the public consumption of statistical results, that leap is made when it shouldn't be. For instance, you can't claim that consumption of ice cream *causes* an increase in murder rates just because they are correlated. In fact, the study showed that temperature was positively correlated with both ice cream sales and murders. (For more on correlation and causation, see Chapter 18.) When can you make the causation leap? The most compelling case is when a well-designed experiment is conducted that rules out other factors that could be related to the outcomes (see Chapter 17 for information on experiments showing cause-and-effect).

You may find yourself wanting to jump to a cause-and-effect relationship when a correlation is found; researchers, the media, and the general public do it all the time. However, before making any conclusions, look at how the data were collected and/or wait to see if other researchers are able to replicate the results (the first thing they try to do after someone else's "groundbreaking result" hits the airwaves).

Part II
Number-Crunching Basics

The 5th Wave   By Rich Tennant

"I ran an evaluation of our last pie chart.
Apparently it's boysenberry."

In this part . . .

Number crunching: It's a dirty job, but somebody has to do it. Why not let it be you? Even if you aren't a numbers person and calculations aren't your thing, the step-by-step approach in this part may be just what you need to boost your confidence in doing and really understanding statistics.

In this part, you get down to the basics of number crunching, from making and interpreting charts and graphs to cranking out and understanding means, medians, standard deviations, and more. You also develop important skills for critiquing someone else's statistical information and getting at the real truth behind the data.

Chapter 5

Means, Medians, and More

*E*very data set has a story, and if statistics are used properly, they do a good job of uncovering and reporting that story. Statistics that are improperly used can tell a different story, or only part of it, so knowing how to make good decisions about the information you're given is very important.

A *descriptive statistic* (or *statistic* for short) is a number that summarizes or describes some characteristic about a set of data. In this chapter, you see some of the most common descriptive statistics and how they are used, and you find out how to calculate them, interpret them, and put them together to get a good picture of a data set. You also find out what these statistics say and what they don't say about the data.

Summing Up Data with Descriptive Statistics

Descriptive statistics take a data set and boil it down to a set of basic information. Summarized data are often used to provide people with information that is easy to understand and that helps answer their questions. Picture your boss coming to you and asking, "What's our client base like these days, and who's buying our products?" How would you like to answer that question — with a long, detailed, and complicated stream of numbers that are sure to glaze her eyes over? Probably not. You want clean, clear, and concise statistics that sum up the client base for her, so that she can see how brilliant you are and then send you off to collect even more data to see how she can include more people in the client base. (That's what you get for being efficient.)

Summarizing data has other purposes, as well. After all the data have been collected from a survey or some other kind of study, the next step is for the researcher to try to make sense out of the data. Typically, the first step researchers take is to run some basic statistics on the data to get a rough idea about what's happening in it. Later in the process, researchers can do more analyses to formulate or test claims made about the population the data came from, estimate certain characteristics about the population (like the mean), look for links between variables they measured, and so on.

Another big part of research is reporting the results, not only to your peers, but also to the media and the general public. Although a researcher's peers may be anxiously waiting to hear about all the complex analyses that were done on a data set, the general public is neither ready for nor interested in that. What does the public want? Basic information. Statistics that make a point clearly and concisely are usually used to relay information to the media and to the public.

If you really need to learn more from data, a quick statistical overview isn't enough. In the statistical world, less is not more, and sometimes the real story behind the data can get lost in the shuffle. To be an informed consumer of statistics, you need to think about which statistics are being reported, what these statistics really mean, and what information is missing. This chapter focuses on these issues.

Crunching Categorical Data: Tables and Percents

Categorical data (also known as *qualitative data*) capture qualities or characteristics about the individual, such as a person's eye color, gender, political party, or opinion on some issue (using categories such as Agree, Disagree, or No opinion). Categorical data tend to fall into groups or categories pretty naturally. "Political party," for example, typically has four groups in the United States: Democrat, Republican, Independent, and Other. Categorical data often come from survey data, but they can also be collected in experiments. For example, in an experimental test of a new medical treatment, researchers may use three categories to assess the outcome of the experiment: Did the patient get better, worse, or stay the same while undergoing the treatment?

Categorical data are often summarized by reporting the percentage of individuals falling into each category. For example, pollsters may report political affiliation statistics by giving the percentage of Republicans, Democrats, Independents, and Others. To calculate the percentage of individuals in a certain category, find the number of individuals in that category, divide by the total number of people in the study, and then multiply by 100%. For example, if a survey of 2,000 teenagers included 1,200 females and 800 males, the resulting percentages would be (1,200 ÷ 2,000) * 100% = 60% female and (800 ÷ 2,000) * 100% = 40% male.

You can break down categorical data further by creating something called two-way tables. *Two-way tables* (also called *crosstabs*) are tables with rows and columns. They summarize the information from two categorical variables at once, such as gender and political party, so you can see (or easily calculate) the percentage of individuals in each combination of categories and use them to make comparisons between groups.

For example, if you had data about the gender and political party of your respondents, you would be able to look at the percentage of Republican females, Republican males, Democratic females, Democratic males, and so on. In this example, the total number of possible combinations in your table would be 2 * 4 = 8, or the total number of gender categories times the total number of party affiliation categories. (See Chapter 19 for the full scoop, and then some, on two-way tables.)

The U.S. government calculates and summarizes loads of categorical data using crosstabs. Typical age and gender data, reported by the U.S. Census Bureau for a survey conducted in 2009, are shown in Table 5-1. (Normally age would be considered a numerical variable, but the way the U.S. government reports it, age is broken down into categories, making it a categorical variable.)

Table 5-1	U.S. Population, Broken Down by Age and Gender (2009)					
Age Group	*Both Sexes*	*%*	*Males*	*%*	*Females*	*%*
Under 5	21,299,656	6.94	10,887,008	7.19	10,412,648	6.69
5–9	20,609,634	6.71	10,535,900	6.96	10,073,734	6.48
10–14	19,973,564	6.51	10,222,522	6.75	9,751,042	6.27
15–19	21,537,837	7.02	11,051,289	7.30	10,486,548	6.74
20–24	21,539,559	7.02	11,093,552	7.32	10,446,007	6.72
25–29	21,677,719	7.06	11,115,560	7.34	10,562,159	6.79
30–34	19,888,603	6.48	10,107,974	6.67	9,780,629	6.29
35–39	20,538,351	6.69	10,353,016	6.84	10,185,335	6.55
40–44	20,991,605	6.84	10,504,139	6.94	10,487,466	6.74

(continued)

Table 5-1 *(continued)*

Age Group	Both Sexes	%	Males	%	Females	%
45–49	22,831,092	7.44	11,295,524	7.46	11,535,568	7.42
50–54	21,761,391	7.09	10,677,847	7.05	11,083,544	7.13
55–59	18,975,026	6.18	9,204,666	6.08	9,770,360	6.28
60–64	15,811,923	5.15	7,576,933	5.00	8,234,990	5.29
65–69	11,784,320	3.84	5,511,164	3.64	6,273,156	4.03
70–74	9,007,747	2.93	4,082,226	2.70	4,925,521	3.17
75–79	7,325,528	2.39	3,149,236	2.08	4,176,292	2.68
80–84	5,822,334	1.90	2,298,260	1.52	3,524,074	2.27
85–89	3,662,397	1.19	1,266,899	0.84	2,395,498	1.54
90–94	1,502,263	0.49	424,882	0.28	1,077,381	0.69
95–99	401,977	0.13	82,135	0.05	319,842	0.21
100+	64,024	0.02	8,758	0.01	55,266	0.04
Total	307,006,550	100.00	151,449,490	100.00	155,557,060	100.00

You can examine many different facets of the U.S. population by looking at and working with different numbers from Table 5-1. For example, looking at gender, you notice that women slightly outnumber men — the population in 2009 was 50.67% female (divide total number of females by total population size and multiply by 100%) and 49.33% male (divide total number of males by total population size and multiply by 100%). You can also look at age: The percentage of the entire population that is under 5 years old was 6.94% (divide the total number under age 5 by the total population size and multiply by 100%). The largest group belongs to the 45–49 year olds, who made up 7.44% of the population.

Next, you can explore a possible relationship between gender and age by comparing various parts of the table. You can compare, for example, the percentage of females to males in the 80-and-over age group. Because these data are reported in 5-year increments, you have to do a little math in order to get your answer, though. The percentage of the population that's female and aged 80 and above (looking at column 7 of Table 5-1) is 2.27% + 1.54% + 0.69% + 0.21% + 0.04% = 4.75%. The percentage of males aged 80 and over (looking at column 5 of Table 5-1) is 1.52% + 0.84% + 0.28% + 0.05% + 0.01% = 2.70%. This shows that the 80-and-over age group for the females is about 76% larger than the males (because [4.75 – 2.70] ÷ 2.70 = 0.76).

These data confirm the widely accepted notion that women tend to live longer than men. However, the gap between men and women is narrowing over time. According to the U.S. Census Bureau, back in 2001 the percentage of women who were 80 years old and over was 4.36, compared to 2.31 for the men. The females in this age group outnumbered the males by a whopping 89% back in 2001 (note that $[4.36 - 2.31] \div 2.31 = 0.89$).

After you have the crosstabs that show the breakdown of two categorical variables, you can conduct hypothesis tests to determine whether a significant relationship or link between the two variables exists, taking into account the fact that data vary from sample to sample. Chapter 14 gives you all the details on hypothesis tests.

Measuring the Center with Mean and Median

With *numerical data,* measurable characteristics such as height, weight, IQ, age, or income are represented by numbers that make sense within the context of the problem (for example in units of feet, dollars, or people). Because the data have numerical meaning, you can summarize them in more ways than is possible with categorical data. The most common way to summarize a numerical data set is to describe where the center is. One way of thinking about what the center of a data set means is to ask, "What's a typical value?" Or, "Where is the middle of the data?" The center of a data set can actually be measured in different ways, and the method chosen can greatly influence the conclusions people make about the data. This section hits on measures of center.

Averaging out to the mean

NBA players make a lot of money, right? You often hear about players like Kobe Bryant or LeBron James who make tens of millions of dollars a year. But is that what the typical NBA player makes? Not really (although I don't exactly feel sorry for the others, given that they still make more money than most of us will ever make). Tens of millions of dollars is the kind of money you can command when you are a superstar among superstars, which is what these elite players are.

So how much money does the typical NBA player make? One way to answer this is to look at the average (the most commonly used statistic of all time).

The *average*, also called the *mean* of a data set, is denoted $\bar{x}$. The formula for finding the mean is:

$$\bar{x} = \frac{\sum x_i}{n}$$

where each value in the data set is denoted by an x with a subscript i that goes from 1 (the first number) to n (the last number).

Here's how you calculate the mean of a data set:

1. **Add up all the numbers in the data set.**

2. **Divide by the number of numbers in the data set, n.**

The mean I discuss here applies to a sample of data and is technically called the *sample mean*. The mean of an entire population of data is denoted with the Greek letter μ and is called the *population mean*. It's found by summing up all the values in the population and dividing by the population size, denoted N (to distinguish it from a sample size, n). Typically the population mean is unknown, and you use a sample mean to estimate it (plus or minus a margin of error; see all the details in Chapter 13).

For example, player salary data for the 13 players on the 2010 NBA Champion Los Angeles Lakers is shown in Table 5-2.

Table 5-2	Salaries for L.A. Lakers NBA Players (2009–2010)
Player	*Salary ($)*
Kobe Bryant	23,034,375
Pau Gasol	16,452,000
Andrew Bynum	12,526,998
Lamar Odom	7,500,000
Ron Artest	5,854,000
Adam Morrison	5,257,229
Derek Fisher	5,048,000
Sasha Vujacic	5,000,000
Luke Walton	4,840,000

Player	Salary ($)
Shannon Brown	2,000,000
Jordan Farmar	1,947,240
Didier Ilunga-Mbenga	959,111
Josh Powell	959,111
Total	**91,378,064**

The mean of all the salaries on this team is $91,378,064 ÷ 13 = $7,029,082. That's a pretty nice average salary, isn't it? But notice that Kobe Bryant really stands out at the top of this list, and he should — his salary was the second highest in the entire league that season (just behind Tracy McGrady). If you remove Kobe from the equation (literally), the average salary of all the Lakers players besides Kobe becomes $68,343,689 ÷ 12 = $5,695,307 — a difference of around 1.3 million.

This new mean is still a hefty amount, but it's significantly lower than the mean salary of all the players including Kobe. (Fans would tell you that this reflects his importance to the team, and others would say no one is worth that much money; this issue is but the tip of the iceberg of the never-ending debates that sports fans — me included — love to have about statistics.)

Bottom line: The mean doesn't always tell the whole story. In some cases it may be a bit misleading, and this is one of those cases. That's because every year a few top-notch players (like Kobe) make much more money than anybody else, and their salaries pull up the overall average salary.

Numbers in a data set that are extremely high or extremely low compared to the rest of the data are called *outliers*. Because of the way the average is calculated, high outliers tend to drive the average upward (as Kobe's salary did in the preceding example). Low outliers tend to drive the average downward.

Splitting your data down the median

Remember in school when you took an exam, and you and most of the rest of the class did badly, but a couple of nerds got 100? Remember how the teacher didn't curve the scores to reflect the poor performance of most of the class? Your teacher was probably using the average, and the average in that case didn't really represent what statisticians might consider the best measure of center for the students' scores.

What can you report, other than the average, to show what the salary of a "typical" NBA player would be or what the test score of a "typical" student in your class was? Another statistic used to measure the center of a data set is called the median. The median is still an unsung hero of statistics in the sense that it isn't used nearly as often as it should be, although people are beginning to report it more nowadays.

The *median* of a data set is the value that lies exactly in the middle when the data have been ordered. It's denoted in different ways; some people use M and some use $\tilde{x}$. Here are the steps for finding the median of a data set:

1. **Order the numbers from smallest to largest.**

2. **If the data set contains an odd number of numbers, choose the one that is exactly in the middle. You've found the median.**

3. **If the data set contains an even number of numbers, take the two numbers that appear in the middle and average them to find the median.**

The salaries for the Los Angeles Lakers during the 2009–2010 season (refer to Table 5-2) are ordered from smallest (at the bottom) to largest (at the top). Because the list contains the names and salaries of 13 players, the middle salary is the seventh one from the bottom: Derek Fisher, who earned $5.048 million that season from the Lakers. Derek is at the median.

This median salary ($5.048 million) is well below the average of $7.029 million for the 2009–2010 Lakers team. Notice that only 4 players of the 13 earned more than the average Lakers salary of $7.029 million. Because the average includes outliers (like the salary of Kobe Bryant), the median salary is more representative of center for the team salaries. The median isn't affected by the salaries of those players who are way out there on the high end the way the average is.

Note: By the way, the lowest Lakers' salary for the 2009–2010 season was $959,111 — a lot of money by most people's standards, but peanuts compared to what you imagine when you think of an NBA player's salary!

The U.S. government most often uses the median to represent the center with respect to income data again because the median is not affected by outliers. For example, the U.S. Census Bureau reported that in 2008, the median household income was $50,233 while the mean was found to be $68,424. That's quite a difference!

Comparing means and medians: Histograms

Sometimes the mean versus median debate can get quite interesting. Suppose you're part of an NBA team trying to negotiate salaries. If you represent the owners, you want to show how much everyone is making and how much money you're spending, so you want to take into account those superstar players and report the average. But if you're on the side of the players, you would want to report the median, because that's more representative of what the players in the middle are making. Fifty percent of the players make a salary above the median, and 50 percent make a salary below the median. To sort it all out, it's best to find and compare both the mean and the median. A graph showing the shape of the data is a great place to start.

One of the graphs you can make to illustrate the shape of numerical data (how many values are close to/far from the mean, where the center is, how many outliers there might be) is a histogram. A *histogram* is a graph that organizes and displays numerical data in picture form, showing groups of data and the number or percentage of the data that fall into each group. It gives you a nice snapshot of the data set. (See Chapter 7 for more information on histograms and other types of data displays.)

Data sets can have many different possible shapes; here is a sampling of three shapes that are commonly discussed in introductory statistics courses:

✔ If most of the data are on the left side of the histogram but a few larger values are on the right, the data are said to be *skewed to the right*.

Histogram A in Figure 5-1 shows an example of data that are skewed to the right. The few larger values bring the mean upwards but don't really affect the median. So when data are skewed right, *the mean is larger than the median*. An example of such data is NBA salaries.

✔ If most of the data are on the right, with a few smaller values showing up on the left side of the histogram, the data are *skewed to the left*.

Histogram B in Figure 5-1 shows an example of data that are skewed to the left. The few smaller values bring the mean down, and again the median is minimally affected (if at all). An example of skewed-left data is the amount of time students use to take an exam; some students leave early, more of them stay later, and many stay until the bitter end (some would stay forever if they could!). When data are skewed left, *the mean is smaller than the median*.

✔ If the data are *symmetric,* they have about the same shape on either side of the middle. In other words, if you fold the histogram in half, it looks about the same on both sides.

Histogram C in Figure 5-1 shows an example of symmetric data in a histogram. With symmetric data, the mean and median are close together.

By looking at Histogram A in Figure 5-1 (whose shape is skewed right), you can see that the "tail" of the graph (where the bars are getting shorter) is to the right, while the "tail" is to the left in Histogram B (whose shape is skewed left). By looking at the direction of the tail of a skewed distribution, you determine the direction of the skewness. Always add the direction when describing a skewed distribution.

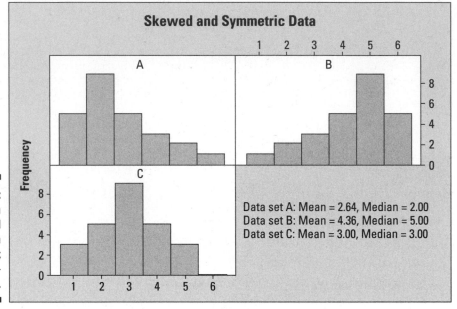

Figure 5-1:
A) Data
skewed
right; B) data
skewed left;
and C) sym-
metric data.

Histogram C is symmetric (it has about the same shape on each side). However, not all symmetric data has a bell shape like Histogram C does. As long as the shape is approximately the same on both sides, then you say that the shape is symmetric.

The average (or mean) of a data set is affected by outliers, but the median is not. In statistical lingo, if a statistic is not affected by a certain characteristic of the data (such as outliers, or skewness), then you say that statistic is *resistant* to that characteristic. In this case the median is resistant to outliers; the mean is not. If someone reports the average value, also ask for the median so that you can compare the two statistics and get a better feel for what's actually going on in the data and what's truly typical.

Accounting for Variation

Variation always exists in a data set, regardless of which characteristics you're measuring, because not every individual is going to have the same exact value for every variable. Variation is what makes the field of statistics what it is. For example, the price of homes varies from house to house, from year to year, and from state to state. The amount of time it takes you to get to work varies from day to day. The trick to dealing with variation is to be able to measure that variation in a way that best captures it.

Reporting the standard deviation

By far the most common measure of variation for numerical data is the standard deviation. The *standard deviation* measures how concentrated the data are around the mean; the more concentrated, the smaller the standard deviation. It's not reported nearly as often as it should be, but when it is, you often see it in parentheses: ($s = 2.68$).

Calculating standard deviation

The formula for the sample standard deviation of a data set (s) is

$$s = \sqrt{\frac{\sum (x - \bar{x})^2}{n-1}}$$

To calculate s, do the following steps:

1. **Find the average of the data set, $\bar{x}$.**

2. **Take each number in the data set (x) and subtract the mean from it to get $(x - \bar{x})$.**

3. **Square each of the differences, $(x - \bar{x})^2$.**

4. **Add up all of the results from Step 3 to get the sum of squares: $\sum (x - \bar{x})^2$.**

5. **Divide the sum of squares (found in Step 4) by the number of numbers in the data set minus one; that is, ($n - 1$). Now you have:**

$$\frac{\sum (x - \bar{x})^2}{n-1}$$

6. **Take the square root to get**

$$s = \sqrt{\frac{\sum (x - \bar{x})^2}{n-1}}$$

   which is the sample standard deviation, s. Whew!

At the end of Step 5 you have found a statistic called the *sample variance,* denoted by s^2. The variance is another way to measure variation in a data set; its downside is that it's in square units. If your data are in dollars, for example, the variance would be in square dollars — which makes no sense. That's why we proceed to Step 6. Standard deviation has the same units as the original data.

Look at the following small example: Suppose you have four quiz scores: 1, 3, 5, and 7. The mean is $16 \div 4 = 4$ points. Subtracting the mean from each number, you get $(1 - 4) = -3$, $(3 - 4) = -1$, $(5 - 4) = +1$, and $(7 - 4) = +3$. Squaring each of these results, you get 9, 1, 1, and 9. Adding these up, the sum is 20. In this example, $n = 4$, and therefore $n - 1 = 3$, so you divide 20 by 3 to get 6.67. The units here are "points squared," which obviously makes no sense. Finally, you take the square root of 6.67, to get 2.58. The standard deviation for these four quiz scores is 2.58 points.

Because calculating the standard deviation involves many steps, in most cases you have a computer calculate it for you. However, knowing how to calculate the standard deviation helps you better interpret this statistic and can help you figure out when the statistic may be wrong.

Statisticians divide by $n - 1$ instead of by n in the formula for s so the results have nicer properties that operate on a theoretical plane that's beyond the scope of this book (not the *Twilight Zone* but close; trust me, that's more than you want to know about *that!*).

The standard deviation *of an entire population of data* is denoted with the Greek letter σ. When I use the term *standard deviation,* I mean s, the sample standard deviation. (When I refer to the population standard deviation, I let you know.)

Interpreting standard deviation

Standard deviation can be difficult to interpret as a single number on its own. Basically, a small standard deviation means that the values in the data set are close to the mean of the data set, on average, and a large standard deviation means that the values in the data set are farther away from the mean, on average.

A small standard deviation can be a goal in certain situations where the results are restricted, for example, in product manufacturing and quality control. A particular type of car part that has to be 2 centimeters in diameter to fit properly had better not have a very big standard deviation during the manufacturing process. A big standard deviation in this case would mean that lots of parts end up in the trash because they don't fit right; either that or the cars will have problems down the road.

But in situations where you just observe and record data, a large standard deviation isn't necessarily a bad thing; it just reflects a large amount of variation in the group that is being studied. For example, if you look at salaries for everyone in a certain company, including everyone from the student intern to the CEO, the standard deviation may be very large. On the other hand, if you narrow the group down by looking only at the student interns, the standard deviation is smaller, because the individuals within this group have salaries that are less variable. The second data set isn't better, it's just less variable.

Similar to the mean, outliers affect the standard deviation (after all, the formula for standard deviation includes the mean). In the NBA salaries example, the salaries of the L.A. Lakers in the 2009–2010 season (shown in Table 5-2) range from the highest, $23,034,375 (Kobe Bryant) down to $959,111 (Didier Ilunga-Mbenga and Josh Powell). Lots of variation, to be sure! The standard deviation of the salaries for this team turns out to be $6,567,405; it's almost as large as the average. However, as you may guess, if you remove Kobe Bryant's salary from the data set, the standard deviation decreases because the remaining salaries are more concentrated around the mean. The standard deviation becomes $4,671,508.

Watch for the units when determining whether a standard deviation is large. For example, a standard deviation of 2 in units of years is equivalent to a standard deviation of 24 in units of months. Also look at the value of the mean when putting standard deviation into perspective. If the average number of Internet newsgroups that a user posts to is 5.2 and the standard deviation is 3.4, that's a lot of variation, relatively speaking. But if you're talking about the age of the newsgroup users where the mean is 25.6 years, that same standard deviation of 3.4 would be comparatively smaller.

Understanding properties of standard deviation

Here are some properties that can help you when interpreting a standard deviation:

- ✔ The standard deviation can never be a negative number, due to the way it's calculated and the fact that it measures a distance (distances are never negative numbers).

- ✔ The smallest possible value for the standard deviation is 0, and that happens only in contrived situations where every single number in the data set is exactly the same (no deviation).

- ✔ The standard deviation is affected by outliers (extremely low or extremely high numbers in the data set). That's because the standard deviation is based on the distance from the *mean*. And remember, the mean is also affected by outliers.

- ✔ The standard deviation has the same units as the original data.

Lobbying for standard deviation

The standard deviation is a commonly used statistic, but it doesn't often get the attention it deserves. Although the mean and median are out there in common sight in the everyday media, you rarely see them accompanied by any measure of how diverse that data set was, and so you are getting only part of the story. In fact, you could be missing the most interesting part of the story.

Without standard deviation, you can't get a handle on whether the data are close to the average (as are the diameters of car parts that come off of a conveyor belt when everything is operating correctly) or whether the data are spread out over a wide range (as are house prices and income levels in the U.S.).

For example if someone told you that the average starting salary for someone working at Company Statistix is $70,000, you may think, "Wow! That's great." But if the standard deviation for starting salaries at Company Statistix is $20,000, that's a lot of variation in terms of how much money you can make, so the average starting salary of $70,000 isn't as informative in the end, is it?

On the other hand, if the standard deviation was only $5,000, you would have a much better idea of what to expect for a starting salary at that company. Which is more appealing? That's a decision each person has to make; however it'll be a much more informed decision once you realize standard deviation matters.

Without the standard deviation, you can't compare two data sets effectively. Suppose two sets of data have the same average; does that mean that the data sets must be exactly the same? Not at all. For example, the data sets 199, 200, 201; and 0, 200, 400 both have the same average (200) yet they have very different standard deviations. The first data set has a *very* small standard deviation ($s=1$) compared to the second data set ($s=200$).

References to the standard deviation may become more commonplace in the media as more and more people (like you, for example) discover what the standard deviation can tell them about a set of results and start asking for it. In your career, you are likely to see the standard deviation reported and used as well.

Being out of range

The range is another statistic that some folks use to measure diversity in a data set. The *range* is the largest value in the data set minus the smallest value in the data set. It's easy to find; all you do is put the numbers in order (from smallest to largest) and do a quick subtraction. Maybe that's why the range is used so often; it certainly isn't because of its interpretative value.

The range of a data set is almost meaningless. It depends on only two numbers in the data set, both of which may reflect extreme values (outliers). My advice is to ignore the range and find the standard deviation, which is a more informative measure of the variation in the data set because it involves all the values. Or you can also calculate another statistic called the *interquartile range,* which is similar to the range with an important difference — it eliminates outlier and skewness issues by only looking at the middle 50% of the data and finding the range for those values. The section "Exploring interquartile range" at the end of this chapter gives you more details.

Examining the Empirical Rule (68-95-99.7)

Putting a measure of center (such as the mean or median) together with a measure of variation (such as standard deviation or interquartile range) is a good way to describe the values in a population. In the case where the data are in the shape of a bell curve (that is, they have a normal distribution; see Chapter 9), the population mean and standard deviation are the combination of choice, and a special rule links them together to get some pretty detailed information about the population as a whole.

The *Empirical Rule* says that if a population has a normal distribution with population mean μ and standard deviation σ, then:

- About 68% of the values lie within 1 standard deviation of the mean (or between the mean minus 1 times the standard deviation, and the mean plus 1 times the standard deviation). In statistical notation, this is represented as $\mu \pm 1\sigma$.

- About 95% of the values lie within 2 standard deviations of the mean (or between the mean minus 2 times the standard deviation, and the mean plus 2 times the standard deviation). The statistical notation for this is $\mu \pm 2\sigma$.

- About 99.7% of the values lie within 3 standard deviations of the mean (or between the mean minus 3 times the standard deviation and the mean plus 3 times the standard deviation). Statisticians use the following notation to represent this: $\mu \pm 3\sigma$.

The Empirical Rule is also known as the *68-95-99.7 Rule,* in correspondence with those three properties. It's used to describe a population rather than a sample, but you can also use it to help you decide whether a sample of data came from a normal distribution. If a sample is large enough and you can see that its histogram looks close to a bell-shape, you can check to see whether the data follow the 68-95-99.7 percent specifications. If yes, it's reasonable to conclude the data came from a normal distribution. This is huge because the normal distribution has lots of perks, as you can see in Chapter 9.

Figure 5-2 illustrates all three components of the Empirical Rule.

The reason that so many (about 68%) of the values lie within 1 standard deviation of the mean in the Empirical Rule is because when the data are bell-shaped, the majority of the values are mounded up in the middle, close to the mean (as Figure 5-2 shows).

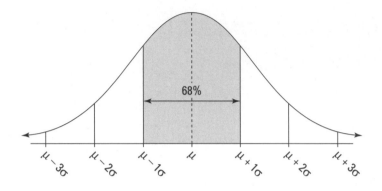

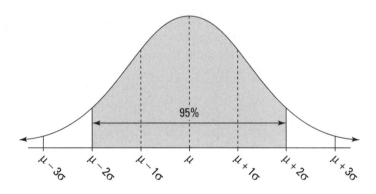

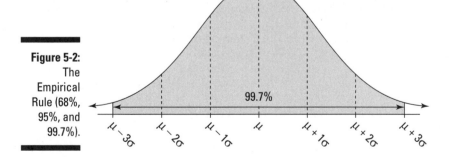

Figure 5-2:
The
Empirical
Rule (68%,
95%, and
99.7%).

Adding another standard deviation on either side of the mean increases the percentage from 68 to 95, which is a big jump and gives a good idea of where "most" of the data are located. Most researchers stay with the 95% range (rather than 99.7%) for reporting their results, because increasing the range to 3 standard deviations on either side of the mean (rather than just 2) doesn't seem worthwhile, just to pick up that last 4.7% of the values.

The Empirical Rule tells you about what percentage of values are within a certain range of the mean, and I need to stress the word *about*. These results are approximations only, and they only apply if the data follow a normal distribution. However, the Empirical Rule is an important result in statistics because the concept of "going out about two standard deviations to get about 95% of the values" is one that you see mentioned often with confidence intervals and hypothesis tests (see Chapters 13 and 14).

Here's an example of using the Empirical Rule to better describe a population whose values have a normal distribution: In a study of how people make friends in cyberspace using newsgroups, the age of the users of an Internet newsgroup was reported to have a mean of 31.65 years, with a standard deviation of 8.61 years. Suppose the data were graphed using a histogram and were found to have a bell-shaped curve similar to what's shown in Figure 5-2.

According to the Empirical Rule, about 68% of the newsgroup users had ages within 1 standard deviation (8.61 years) of the mean (31.65 years). So about 68% of the users were between ages 31.65 – 8.61 years and 31.65 + 8.61 years, or between 23.04 and 40.26 years. About 95% of the newsgroup users were between the ages of 31.65 – 2(8.61), and 31.65 + 2(8.61), or between 14.43 and 48.87 years. Finally, about 99.7% of the newsgroup users' ages were between 31.65 – 3(8.61) and 31.65 + 3(8.61), or between 5.82 and 57.48 years.

This application of the rule gives you a much better idea about what's happening in this data set than just looking at the mean, doesn't it? As you can see, the mean and standard deviation used together add value to your results; plugging these values into the Empirical Rule allows you to report ranges for "most" of the data yourself.

Remember, the condition for being able to use the Empirical Rule is that the data have a normal distribution. If that's not the case (or if you don't know what the shape actually is), you can't use it. To describe your data in these cases, you can use percentiles, which represent certain cutoff points in the data (see the later section "Gathering a five-number summary").

Measuring Relative Standing with Percentiles

Sometimes the precise values of the mean, median, and standard deviation just don't matter, and all you are interested in is where you stand compared to the rest of the herd. In this situation, you need a statistic that reports *relative standing*, and that statistic is called a percentile. The kth *percentile* is a number in the data set that splits the data into two pieces: The lower piece contains k percent of the data, and the upper piece contains the rest of the data (which amounts to [100 – k] percent, because the total amount of data is 100%). *Note:* k is any number between 1 and 100.

The median is the 50th percentile: The point in the data where 50% of the data fall below that point, and 50% fall above it.

In this section, you find out how to calculate, interpret, and put together percentiles to help you uncover the story behind a data set.

Calculating percentiles

To calculate the kth percentile (where k is any number between one and one hundred), do the following steps:

1. **Order all the numbers in the data set from smallest to largest.**

2. **Multiply k percent times the total number of numbers, n.**

3a. **If your result from Step 2 is a whole number, go to Step 4. If the result from Step 2 is not a whole number, round it up to the nearest whole number and go to Step 3b.**

3b. **Count the numbers in your data set from left to right (from the smallest to the largest number) until you reach the value indicated by Step 3a. The corresponding value in your data set is the kth percentile.**

4. **Count the numbers in your data set from left to right until you reach the one indicated by Step 2. The kth percentile is the average of that corresponding value in your data set and the value that directly follows it.**

For example, suppose you have 25 test scores, and in order from lowest to highest they look like this: 43, 54, 56, 61, 62, 66, 68, 69, 69, 70, 71, 72, 77, 78, 79, 85, 87, 88, 89, 93, 95, 96, 98, 99, 99. To find the 90th percentile for these (ordered) scores, start by multiplying 90% times the total number of scores, which gives 90% * 25 = 0.90 * 25 = 22.5. Rounding up to the nearest whole number, you get 23.

Counting from left to right (from the smallest to the largest number in the data set), you go until you find the 23rd number in the data set. That number is 98, and it's the 90th percentile for this data set.

Now say you want to find the 20th percentile. Start by taking $0.20 * 25 = 5$; this is a whole number, so proceed from Step 3a to Step 4, which tells us the 20th percentile is the average of the 5th and 6th numbers in the ordered data set (62 and 66). The 20th percentile then comes to $(62 + 66) \div 2 = 64$. The median (the 50th percentile) for the test scores is the 13th score: 77.

There is no single definitive formula for calculating percentiles. The formula here is designed to make finding the percentile easier and more intuitive, especially if you're doing the work by hand; however, other formulas are used when you're working with technology. The results you get using various methods may differ but not by much.

Interpreting percentiles

Percentiles report the relative standing of a particular value within a data set. If that's what you're most interested in, the actual mean and standard deviation of the data set are not important, and neither is the actual data value. What's important is where you stand — not in relation to the mean, but in relation to everyone else: That's what a percentile gives you.

For example, In the case of exam scores, who cares what the mean is, as long as you scored better than most of the class? Who knows, it may have been an impossible exam and 40 points out of 100 was a great score (that happened to me in an advanced math class once; heaven forbid this should ever happen to you!). In this case, your score itself is meaningless, but your percentile tells you everything.

Suppose your exam score is better than 90% of the rest of the class. That means your exam score is at the 90th percentile (so $k = 90$), which hopefully gets you an A. Conversely, if your score is at the 10th percentile (which would never happen to you, because you're such an excellent student), then $k = 10$; that means only 10% of the other scores are below yours, and 90% of them are above yours; in this case an A is not in your future.

A nice property of percentiles is they have a universal interpretation: Being at the 95th percentile means the same thing no matter if you are looking at exam scores or weights of packages sent through the postal service; the 95th percentile always means 95% of the other values lie below yours, and 5% lie above it. This also allows you to fairly compare two data sets that have different means and standard deviations (like ACT scores in reading versus math). It evens the playing field and gives you a way to compare apples to oranges, so to speak.

A percentile is *not* a percent; a percentile is a number (or the average of two numbers) in the data set that marks a certain percentage of the way through the data. Suppose your score on the GRE was reported to be the 80th percentile. This doesn't mean you scored 80% of the questions correctly. It means that 80% of the students' scores were lower than yours and 20% of the students' scores were higher than yours.

A high percentile doesn't always constitute a good thing. For example, if your city is at the 90th percentile in terms of crime rate compared to cities of the same size, that means 90% of cities similar to yours have a crime rate that is lower than yours, which is not good for you. Another example is golf scores; a low score in golf is a good thing, so being at the 80th percentile with your score wouldn't qualify you for the PGA tour, let's just say that.

Comparing household incomes

The U.S. government often reports percentiles among its data summaries. For example, the U.S. Census Bureau reported the median (the 50th percentile) household income for 2001 to be $42,228, and in 2007 it was reported to be $50,233. The Bureau also reports various percentiles for household income for each year, including the 10th, 20th, 50th, 80th, 90th, and 95th. Table 5-3 shows the values of each of these percentiles for both 2001 and 2007.

Table 5-3	U.S. Household Income (2001 versus 2007)	
Percentile	*2001 Household Income*	*2007 Household Income*
10th	$10,913	$12,162
20th	$17,970	$20,291
50th	$42,228	$50,233
80th	$83,500	$100,000
90th	$116,105	$136,000
95th	$150,499	$177,000

Looking at the percentiles for 2001 in Table 5-3, you can see that the bottom half of the incomes are closer together than the top half of the incomes are. The difference between the 20th percentile and the 50th percentile is about $24,000, whereas the spread between the 50th percentile and the 80th percentile is more like $41,000. The difference between the 10th and 50th percentiles is only about $31,000, whereas the difference between the 50th and the 90th percentiles is a whopping $74,000.

The percentiles for 2007 are all higher than the percentiles for 2001 (which is a good thing!). They are also more spread out. For 2007, the difference between the 20th and 50th percentiles is around $30,000, and from the 50th to the 80th it's approximately $50,000; both of these differences are larger than for 2001. Similarly, the 10th percentile is farther from the 50th (about $38,000 difference) in 2007 compared to 2001, and the 50th is farther from the 90th (by about $86,000) in 2007, compared to 2001. These results tell us that incomes are increasing in general at all levels between 2001 and 2007, but the gap is widening between those levels. For example, the 10th percentile for income in 2001 was $10,913 (as seen in Table 5-3), compared to $12,162 in 2007; this represents about an 11 percent increase (subtract the two and divide by 10,913). Now compare the 95th percentiles for 2007 versus 2001; the increase is almost 18%. Now, technically, you may want to adjust the 2001 values for inflation, but you get the basic idea.

Percentage changes affect the variability in a data set. For example, when salary raises are given on a percentage basis, the diversity in the salaries also increases; it's the "rich get richer" idea. The guy making $30,000 gets a 10 percent raise and his salary goes up to $33,000 (an increase of $3,000); but the guy making $300,000 gets a 10 percent raise and now makes $330,000 (a difference of $30,000). So when you first get hired for a new job, negotiate the highest possible salary you can because your raises that follow will also net a higher amount.

Examining ACT Scores

Each year millions of U.S. high school students take a nationally administered ACT exam as part of the process of applying for colleges. The test is designed to assess college readiness in the areas of English, Math, Reading, and Science. Each test has a possible score of 36 points.

ACT does not release the average or standard deviation of the test scores for a given exam. (That would be a real hassle if they did, because these statistics can change from exam to exam, and people would complain that this exam was harder than that exam when the actual scores are not relevant.) To avoid these issues, and for other reasons, ACT reports test results using percentiles.

Percentiles are usually reported in the form of a predetermined list. For example, the U.S. Census Bureau reports the 10th, 20th, 50th, 80th, 90th, and 95th percentiles for household income (as shown in Table 5-3). However, ACT uses percentiles in a different way. Rather than reporting the exam scores corresponding to a premade list of percentiles, they list each possible exam score and report its corresponding percentile, whatever that turns out to be. That way, to find out where you stand, you just look up your score and you'll find out your percentile.

Table 5-4 shows the 2009 percentiles for the scores on the Mathematics and Reading ACT exams. To interpret an exam score, find the row corresponding to the score and the column for the exam area (for example, Reading). Intersect row and column and you find out which percentile your score represents; in other words, you see what percentage of your fellow exam-taking comrades scored lower than you.

Table 5-4	Percentiles for All Possible ACT Exam Scores in Math and Reading	
ACT Score	*Mathematics Percentile*	*Reading Percentile*
34–36	99	99
33	98	97
32	97	95
31	96	93
30	95	91
29	93	88
28	91	85
27	88	81
26	84	78
25	79	74
24	74	70
23	68	65
22	62	59
21	57	54
20	52	47
19	47	41
18	40	34
17	33	30
16	24	24
15	14	19
14	06	14
13	02	09
12	01	06
11	01	03
1–10	01	01

For example, suppose you scored 30 on the Math exam; in Table 5-4 you look at the row for 30 in the column for Math; you see your score is at the 95th percentile. In other words 95% of the students scored lower than you, and only 5% scored higher than you.

Now suppose you also scored a 30 on the Reading exam. Just because a score of 30 represents the 95th percentile for Math doesn't necessarily mean a score of 30 is at the 95th percentile for Reading as well. (It's probably reasonable to expect that fewer people score 30 or higher on the Math exam than on the Reading exam.)

To test my theory, look at column 3 of Table 5-4 in the row for a score of 30. You see that a score of 30 on the Reading exam puts you at the 91st percentile — not quite as great as your position on the Math exam, but certainly not a bad score.

Gathering a five-number summary

Beyond reporting a single measure of center and/or a single measure of spread, you can create a group of statistics and put them together to get a more detailed description of a data set. The Empirical Rule (as seen in "Examining the Empirical Rule (68-95-99.7)" earlier in this chapter) uses the mean and standard deviation in tandem to describe a bell-shaped data set. In the case where your data are not bell-shaped, you use a different set of statistics (based on percentiles) to describe the big picture of your data. This method involves cutting the data into four pieces (with an equal amount of data in each piece) and reporting the resulting five cutoff points that separate these pieces. These cutoff points are represented by a set of five statistics that describe how the data are laid out.

The *five-number summary* is a set of five descriptive statistics that divide the data set into four equal sections. The five numbers in a five-number summary are:

1. The *minimum* (smallest) number in the data set

2. The *25th percentile* (also known as *the first quartile,* or Q_1)

3. The *median* (50th percentile)

4. The *75th percentile* (also known as *the third quartile,* or Q_3)

5. The *maximum* (largest) number in the data set

For example, suppose you want to find the five-number summary of the following 25 (ordered) exam scores: 43, 54, 56, 61, 62, 66, 68, 69, 69, 70, 71, 72, 77, 78, 79, 85, 87, 88, 89, 93, 95, 96, 98, 99, 99. The minimum is 43, the maximum is 99, and the median is the number directly in the middle, 77.

To find Q_1 and Q_3 you use the steps shown in the section "Calculating percentiles," with $n = 25$. Step 1 is done because the data are ordered. For Step 2, since Q_1 is the 25th percentile, multiply $0.25 * 25 = 6.25$. This is not a whole number, so Step 3a says to round it up to 7 and proceed to Step 3b.

Following Step 3b, you count from left to right in the data set until you reach the 7th number, 68; this is Q_1. For Q_3 (the 75th percentile) you multiply $0.75 * 25 = 18.75$, which you round up to 19. The 19th number on the list is 89, so that's Q_3. Putting it all together, the five-number summary for these 25 test scores is 43, 68, 77, 89, and 99. To best interpret a five-number summary, you can use a boxplot; see Chapter 7 for details.

Exploring interquartile range

The purpose of the five-number summary is to give descriptive statistics for center, variation, and relative standing all in one shot. The measure of center in the five-number summary is the median, and the first quartile, median, and third quartiles are measures of relative standing.

To obtain a measure of variation based on the five-number summary, you can find what's called the *interquartile range* (or *IQR*). The *IQR* equals $Q_3 - Q_1$ (that is, the 75th percentile minus the 25th percentile) and reflects the distance taken up by the innermost 50% of the data. If the *IQR* is small, you know a lot of data are close to the median. If the *IQR* is large, you know the data are more spread out from the median. The *IQR* for the test scores data set is $89 - 68 = 21$, which is fairly large, seeing as how test scores only go from 0 to 100.

The interquartile range is a much better measure of variation than the regular range (maximum value minus minimum value; see the section "Being out of range" earlier in this chapter). That's because the interquartile range doesn't take outliers into account; it cuts them out of the data set by only focusing on the distance within the middle 50 percent of the data (that is, between the 25th and 75th percentiles).

Descriptive statistics that are well chosen and used correctly can tell you a great deal about a data set, such as where the center is located, how diverse the data are, and where a good portion of the data lies. However, descriptive statistics can't tell you everything about the data, and in some cases they can be misleading. Be on the lookout for situations where a different statistic would be more appropriate (for example, the median describes center more fairly than the mean when the data is skewed), and keep your eyes peeled for situations where critical statistics are missing (for example, when a mean is reported without a corresponding standard deviation).

Chapter 6

Getting the Picture: Graphing Categorical Data

· ·

· ·

Data displays, especially charts and graphs, seem to be everywhere, showing everything from election results, broken down by every conceivable characteristic, to how the stock market has fared over the past few years (months, weeks, days, minutes). We're living in an instant gratification, fast-information society; everyone wants to know the bottom line and be spared the details.

The abundance of graphs and charts is not necessarily a bad thing, but you have to be careful; some of them are incorrect or even misleading (sometimes intentionally and sometimes by accident), and you have to know what to look for.

This chapter is about graphs involving *categorical data* (data that places individuals into groups or categories, such as gender, opinion, or whether a patient takes medication every day. Here you find out how to read and make sense of these data displays and get some tips for evaluating them and spotting problems. (***Note:*** Data displays for *numerical data,* such as weight, exam score, or the *number* of pills taken by a patient each day, come in Chapter 7.)

The most common types of data displays for categorical data are pie charts and bar graphs. In this chapter, I present examples of each type of data display and share some thoughts on interpretation and tips for critically evaluating each type.

Take Another Little Piece of My Pie Chart

A pie chart takes categorical data and breaks them down by group, showing the percentage of individuals that fall into each group. Because a pie chart takes on the shape of a circle, the "slices" that represent each group can easily be compared and contrasted.

Because each individual in the study falls into one and only one category, the sum of all the slices of the pie should be 100% or close to it (subject to a bit of rounding off). However, just in case, keep your eyes open for pie charts whose percentages just don't add up.

Tallying personal expenses

When you spend your money, what do you spend it on? What are your top three expenses? According to the U.S. Bureau of Labor Statistics 2008 Consumer Expenditure Survey, the top six sources of consumer expenditures in the U.S. were housing (33.9%), transportation (17.0%), food (12.8%), personal insurance and pensions (11.1%), healthcare (5.9%), and entertainment (5.6%). These six categories make up over 85% of average consumer expenses. (Although the exact percentages change from year to year, the list of the top six items remains the same.)

Figure 6-1 summarizes the 2008 U.S. expenditures in a pie chart. Notice that the "Other" category is a bit large in this chart (13.7%). However, with so many other possible expenditures out there (including this book), each one would only get a tiny slice of the pie for itself, and the resulting pie chart would be a mess. In this case, it is too difficult to break "Other" down further. (But in many other cases you can.)

Ideally, a pie chart shouldn't have too many slices because a large number of slices distracts the reader from the main point(s) the pie chart is trying to relay. However, lumping the remaining categories into one slice that's one of the largest in the whole pie chart leaves readers wondering what's included in that particular slice. With charts and graphs, doing it right is a delicate balance.

Bringing in a lotto revenue

State lotteries bring in a great deal of revenue, and they also return a large portion of the money received, with some of the revenues going to prizes and some being allocated to state programs such as education. Where does lottery revenue come from? Figure 6-2 is a pie chart showing the types of games and their percentage of revenue as recently reported by Ohio's state lottery. (Note the slices don't sum to 100% exactly due to slight rounding error.)

You can see by the pie chart in Figure 6-2 that 49.3% of the lottery sales revenue comes from the instant (scratch-off) games. The rest come from various lottery-type games in which players choose a set of numbers and win if a certain number of their numbers match those chosen by the lottery.

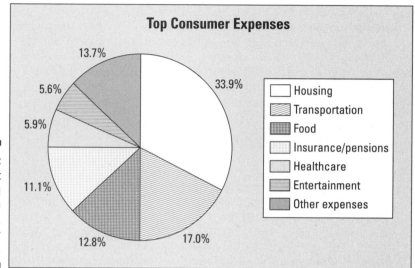

Figure 6-1:
Pie chart
showing
how people
in the U.S.
spend their
money.

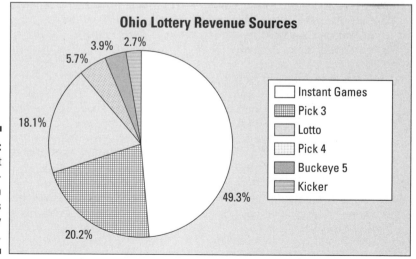

Figure 6-2:
Pie chart
break-
ing down
a state's
lottery
revenue.

Notice that this pie chart doesn't tell you *how much* money came in, only *what percentage* of the money came from each type of game. About half the money (49.3%) came from instant scratch-off games; does this revenue represent a million dollars, two million dollars, ten million dollars, or more? You can't answer these questions without knowing the total amount of revenue dollars.

I was, however, able to find this information on another chart provided by the lottery Web site: The total revenue (over a 10-year period) was reported as "1,983.1 million dollars" — which you also know as 1.9831 billion dollars. Because 49.3% of sales came from instant games, they therefore represent sales revenue of $977,668,300 over a 10-year period. That's a lot of (or dare I say a "lotto") scratching.

Ordering takeout

It's also important to watch for totals when examining a pie chart from a survey. A newspaper I read reported the latest results of a "people poll." They asked, "What is your favorite night to order takeout for dinner?" The results are shown in a pie chart (see Figure 6-3).

You can clearly see that Friday night is the most popular night for ordering takeout (and that result makes sense) with decreasing demand moving from Saturday through Monday. The actual percentages shown in Figure 6-3 really only apply to the people who were surveyed; how close these results mimic the population depends on many factors, one of which is sample size. But unfortunately, sample size is not included as part of this graph. (For example, it would be nice to see "*n* = XXX" below the title; where *n* represents sample size.)

Without knowing the sample size, you can't tell how accurate the information is. Which results would you find to be more accurate — those based on 25 people, 250 people, or 2,500 people? When you see the number 10%, you don't know if it's 10 out of 100, 100 out of 1,000, or even 1 out of 10. To statisticians, $1 \div 10$ is not the same as $100 \div 1,000$, even though they both represent 10%. (Don't tell that to mathematicians — they'll think you're nuts!)

Pie charts often don't include mention of the total sample size. Always check for the sample size, especially if the results are very important to you; don't assume it's large! If you don't see the sample size, go to the source of the data and ask for it.

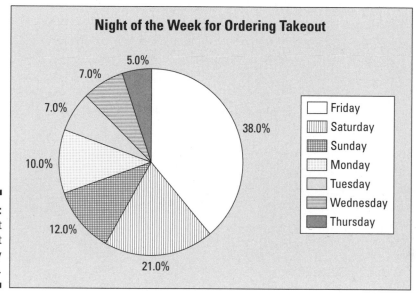

Figure 6-3:
Pie chart
for takeout
food survey
results.

Projecting age trends

The U.S. Census Bureau provides an almost unlimited amount of data, statistics, and graphics about the U.S. population, including the past, present, and projections for the future. It often makes comparisons between years in order to look for changes and trends.

One recent Census Bureau population report looked at what it calls the "older U.S. population" (by the government's definition, this means people 65 years old or over). Age was broken into the following groups: 65–69 years, 70–74 years, 75–79 years, 80–84 years, and 85 and over. The Bureau calculated and reported the percentage in each age group for the year 2010 and made projections for the percentage in each age group for the year 2050.

I made side-by-side pie charts for the years 2010 versus 2050 (projections) to make comparisons; you can see the results in Figure 6-4. The percentage of the older population in each age group for 2010 is shown in one pie chart, and alongside it is a pie chart of the projected percentage for each age group for 2050 (based on the current age of the entire U.S. population, birth and death rates, and other variables).

If you compare the sizes of the slices from one graph to the other in Figure 6-4, you see that the slices for corresponding age groups are larger for the 2050 projections (compared to 2010) as the age groups get older, and the slices are smaller for the 2050 projections (compared to 2010) as the age groups get younger. For example the 65–69 age group decreases from 30% in 2010 to a projected 25% in 2050; while the 85-and-over age group increases from 14% in 2010 to 19% projected for 2050.

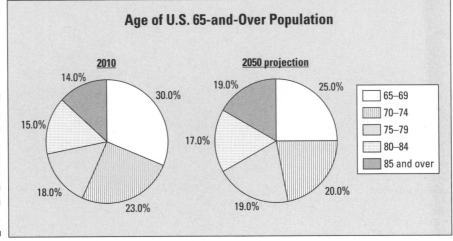

Figure 6-4: Side-by-side pie charts on the aging population, 2010 versus 2050 projections.

The results from Figure 6-4 indicate a shift in the ages of the population toward the older categories. From there, the medical and social research communities can examine the ramifications of this trend in terms of health-care, assisted living, social security, and so on.

The operative words here are *if the trend continues.* As you know, many variables affect population size, and you need to take those into account when interpreting these projections into the future. The U.S. government always points out caveats like this in their reports; it is very diligent about that.

The pie charts in Figure 6-4 work well for comparing groups because they are side-by-side on the same graph, using the same coding for the age groups in each, and their slices are in the same order for both as you move clockwise around the graphs. They aren't all scrambled up on each graph so you have to hunt for a certain age group on each graph separately.

Evaluating a pie chart

The following tips help you taste test a pie chart for statistical correctness:

- ✔ Check to be sure the percentages add up to 100% or very close to it (any round-off error should be very small).

- ✔ Beware of slices of the pie called "Other" that are larger than many of the other slices.

- ✔ Look for a reported total number of units (people, dollar amounts, and so on) so that you can determine (in essence) how "big" the pie was before being divided up into the slices that you're looking at.

- ✔ Avoid three-dimensional pie charts; they don't show the slices in their proper proportions. The slices in front look larger than they should.

Raising the Bar on Bar Graphs

A *bar graph* (or *bar chart*) is perhaps the most common data display used by the media. Like a pie chart, a bar graph breaks categorical data down by group. Unlike a pie chart, it represents these amounts by using bars of different lengths; whereas a pie chart most often reports the amount in each group as percentages, a bar graph uses either the number of individuals in each group (also called the *frequency*) or the percentage in each group (called the *relative frequency*).

Tracking transportation expenses

How much of their income do people in the United States spend on transportation to get back and forth to work? It depends on how much money they make. The Bureau of Transportation Statistics (did you know such a department existed?) conducted a study on transportation in the U.S. recently, and many of its findings are presented as bar graphs like the one shown in Figure 6-5.

This particular bar graph shows how much money is spent on transportation for people in different household-income groups. It appears that as household income increases, the total expenditures on transportation also increase. This makes sense, because the more money people have, the more they have available to spend.

But would the bar graph change if you looked at transportation expenditures not in terms of total dollar amounts, but as the percentage of household income? The households in the first group make less than $5,000 a year and have to spend $2,500 of it on transportation. (**Note:** The label reads "2.5," but because the units are in thousands of dollars, the 2.5 translates into $2,500.)

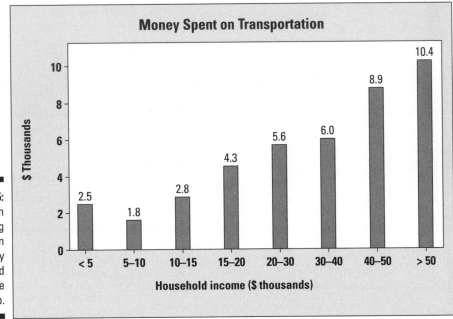

Figure 6-5: Bar graph showing transportation expenses by household income group.

This $2,500 represents 50% of the annual income of those who make $5,000 per year; the percentage of the total income is even higher for those who make less than $5,000 per year. The households earning $30,000–$40,000 per year pay $6,000 per year on transportation, which is between 15% and 20% of their household income. So, although the people making more money spend more dollars on transportation, they don't spend more as a percentage of their total income. Depending on how you look at expenditures, the bar graph can tell two somewhat different stories.

Another point to check out is the groupings on the graph. The categories for household income as shown aren't equivalent. For example, each of the first four bars represents household incomes in intervals of $5,000, but the next three groups increase by $10,000 each, and the last group contains every household making more than $50,000 per year. Bar graphs using different-sized intervals to represent numerical values (such as Figure 6-5) make true comparisons between groups more difficult. (However, I'm sure the government has its reasons for reporting the numbers this way; for example, this may be the way income is broken down for tax-related purposes.)

One last thing: Notice that the numerical groupings in Figure 6-5 overlap on the boundaries. For example, $30,000 appears in both the 5th and 6th bars of the graph. So, if you have a household income of $30,000, which bar do you fall into? (You can't tell from Figure 6-5, but I'm sure the instructions are buried in a huge report in the basement of some building in Washington, D.C.) This kind of overlap appears quite frequently in graphs, but you need to know how the borderline values are being treated. For example, the rule may be "Any data lying exactly on a boundary value automatically goes into the bar to its immediate right." (Looking at Figure 6-5, that puts a household with a $30,000 income into the 6th bar rather than the 5th.) As long as they are being consistent for each boundary, that's okay. The alternative, describing the income boundaries for the 5th bar as "20,000 to $29,999.99," is not an improvement. Along those lines, income data can also be presented using a histogram (see Chapter 7), which has a slightly different look to it.

Making a lotto profit

That lotteries rake in the bucks is a well-known fact; but they also shell it out. How does it all shake out in terms of profits? Figure 6-6 shows the recent sales and expenditures of a certain state lottery.

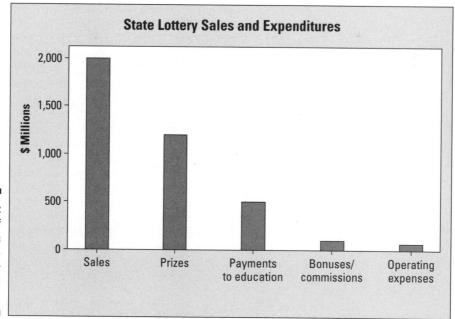

Figure 6-6: Bar graph of lottery sales and expenditures for a certain state.

In my opinion, this bar graph needs some additional info from behind the scenes to make it more understandable. The bars in Figure 6-6 don't represent similar types of entities. The first bar represents sales (a form of revenue), and the other bars represent expenditures. The graph would be much clearer if the first bar weren't included; for example, the total sales could be listed as a footnote.

Tipping the scales on a bar graph

Another way a graph can be misleading is through its choice of scale on the frequency/relative frequency axis (that is, the axis where the amounts in each group are reported), and/or its starting value.

By using a "stretched out" scale (for example, having each half inch of a bar represent 10 units versus 50 units), you can stretch the truth, make differences look more dramatic, or exaggerate values. Truth-stretching can also occur if the frequency axis starts out at a number that's very close to where the differences in the heights of the bars start; you are in essence chopping off the bottom of the bars (the less exciting part) and just showing their tops; emphasizing (in a misleading way) where the action is. Not every frequency axis has to start at zero, but watch for situations that elevate the differences.

A good example of a graph with a stretched out scale is seen in Chapter 3, regarding the results of numbers drawn in the "Pick 3" lottery. (You choose three one-digit numbers and if they all match what's drawn, you win.) In Chapter 3, the percentage of times each number (from 0–9) was drawn is shown in Table 3-2, and the results are displayed in a bar graph in Figure 3-1a. The scale on the graph is stretched and starts at 465, making the differences in the results look larger than they really are; for example, it looks like the number 1 was drawn much less often, whereas the number 2 was drawn much more often, when in reality there is no statistical difference between the percentage of times each number was drawn. (I checked.)

Why was the graph in Figure 3-1a made this way? It might lead people to think they've got an inside edge if they choose the number 2 because it's "on a hot streak"; or they might be led to choose the number 1 because it's "due to come up." Both of these theories are wrong, by the way; because the numbers are chosen at random, what happened in the past doesn't matter. In Figure 3-1b you see a graph that's been made correctly. (For more examples of where our intuition can go wrong with probability and what the scoop really is, see another of my books, *Probability For Dummies*, also published by Wiley.)

Alternatively, by using a "squeezed down" scale (for example, having each half inch of a bar represent 50 units versus 10 units), you can downplay differences, making results look less dramatic than they actually are. For example, maybe a politician doesn't want to draw attention to a big increase in crime from the beginning to the end of her term, so she may have the number of crimes of each type shown where each half inch of a bar represents 500 crimes, versus 100 crimes. This squeezes the numbers together and makes differences less noticeable. Her opponent in the next election would go the other way and use a stretched-out scale to emphasize a crime increase in dramatic fashion, and voilà! (Now you know the answer to the question "How can two people talk about the same data and get two different conclusions?" Welcome to the world of politics.)

With a pie chart, however, the scale can't be changed to over-emphasize (or downplay) the results. No matter how you slice up a pie chart, you're always slicing up a circle, and the proportion of the total pie belonging to any given slice won't change, even if you make the pie bigger or smaller.

Pondering pet peeves

A recent survey of 100 people with office jobs asked them to report their biggest pet peeves in the workplace. (Before going on, you may want to jot down a couple of yours, just for fun.) A bar graph of the results of the survey is shown in Figure 6-7. Poor time management looks to be the number-one issue for these workers (I hope they didn't do this survey on company time).

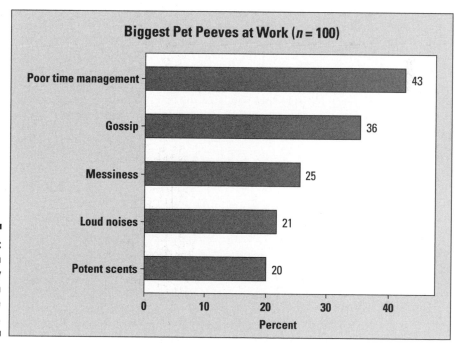

Figure 6-7: Bar graph for survey data with multiple responses.

Evaluating a bar graph

To raise the statistical bar on bar graphs, check out these tips:

✔ Bars that divide up values of a numerical variable (such as income) should be equal in width (if possible) for fair comparison.

✔ Be aware of the scale of the bar graph and determine whether it's an appropriate representation of the information.

✔ Some bar graphs don't sum to one because they are showing the results of more than one variable; make sure it's clear what's being summarized.

✔ Check whether the results are shown as the percentage within each group (relative frequencies) or the number in each group (frequencies).

✔ If you see relative frequencies, check for the total sample size — it matters. If you see frequencies, divide each one by the total sample size to get percentages, which are easier to compare.

If you take a look at the percentages shown for each pet peeve listed, you see they don't sum to one. That tells you that each person surveyed was allowed to choose more than one pet peeve (like that would be hard to do); perhaps they were asked to name their top three pet peeves, for example. For this data set and others like it that allow for multiple responses, a pie chart wouldn't be possible (unless you made one for every single pet peeve on the list).

Note that Figure 6-7 is a *horizontal bar graph* (its bars go side to side) as opposed to a *vertical bar graph* (in which bars go up and down, as in Figure 6-6). Either orientation is fine; use whichever one you prefer when you make a bar graph. Do, however, make sure that you label the axes appropriately and include proper units (such as gender, opinion, or day of the week) where appropriate.

Chapter 7

Going by the Numbers: Graphing Numerical Data

T he main purpose of charts and graphs is to summarize data and display the results to make your point clearly, effectively, and correctly. In this chapter, I present data displays used to summarize *numerical* data — data that represent *counts* (such as the number of pills a patient with diabetes takes per day, or the number of accidents at an intersection per year) or *measurements* (the time it takes you to get to work/school each day, or your blood pressure).

You see examples of how to make, interpret, and evaluate the most common data displays for numerical data: time charts, histograms, and boxplots. I also point out many potential problems that can occur in these graphs, including how people often misread what's there. This information will help you develop important detective skills for quickly spotting misleading graphs.

Handling Histograms

A histogram provides a snapshot of all the data broken down into numerically ordered groups, making it a quick way to get the big picture of the data, in particular, its general shape. In this section you find out how to make and interpret histograms, and how to critique them for correctness and fairness.

Making a histogram

A *histogram* is a special graph applied to data broken down into numerically ordered groups; for example, age groups such as 10–20, 21–30, 31–40, and so on. The bars connect to each other in a histogram — as opposed to a bar graph (Chapter 6) for categorical data, where the bars represent categories that don't have a particular order, and are separated. The height of each bar of a histogram represents either the number of individuals (called the *frequency*) in each group or the percentage of individuals (the *relative frequency*) in each group. Each individual in the data set falls into exactly one bar.

You can make a histogram from any numerical data set; however, you can't determine the actual values of the data set from a histogram because all you know is which group each data value falls into.

An award winning example

Here's an example of how to create a histogram for all you movie lovers out there (especially those who love old movies). The Academy Awards started in 1928, and one of the most popular categories for this award is Best Actress in a Motion Picture. Table 7-1 shows the winners of the first eight Best Actress Oscars, the years they won (1928–1935), their ages at the time of winning their awards, and the movies they were in. From the table you see the ages range from 22 to 62 — much wider than you may have thought it would be.

Table 7-1	Ages of Best Actress Oscar Award Winners 1928–1935		
Year	**Winner**	**Age**	**Movie**
1928	Laura Gainor	22	*Sunrise*
1929	Mary Pickford	37	*Coquette*
1930	Norma Shearer	30	*The Divorcee*
1931	Marie Dressler	62	*Min and Bill*
1932	Helen Hayes	32	*The Sin of Madelon Claudet*
1933	Katharine Hepburn	26	*Morning Glory*
1934	Collette Colbert	31	*It Happened One Night*
1935	Bette Davis	27	*Dangerous*

To find out more about the ages of Best Actresses, I expanded my data set to the period 1928–2009. The age variable for this data set is numerical, so you can graph it using a histogram. From there you can answer questions like: What do the ages of these actresses look like? Are they mostly young, old, in between? Are their ages all spread out, or are they similar? Are most of them in a certain age range, with a few outliers (either very young or very old actresses, compared to the others)? To investigate these questions, a histogram of ages of the Best Award actresses is shown in Figure 7-1.

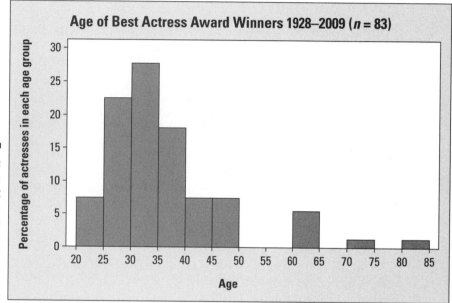

Figure 7-1:
Histogram
of Best
Actress
Academy
Award
winners'
ages,
1928–2009.

Notice that the age groups are shown on the horizontal (*x*) axis. They go by groups of 5 years each: 20–25, 25–30, 30–35, . . . 80–85. The percentage (relative frequency) of actresses in each age group appears on the vertical (*y*) axis. For example, about 27 percent of the actresses were between 30 and 35 years of age when they won their Oscars.

Creating appropriate groups

For Figure 7-1, I used groups of 5 years each in the above example because increments of 5 create natural breaks for years and because it provides enough bars to look for general patterns. You don't have to use this particular grouping, however; you have a bit of poetic license when making a histogram. (However, this freedom allows others to deceive you as you see in the later section "Detecting misleading histograms.") Here are some tips for setting up your histogram:

✔ Each data set requires different ranges for its groupings, but you want to avoid ranges that are too wide or too narrow.

• If a histogram has really wide ranges for its groups, it places all the data into a very small number of bars that make meaningful comparisons impossible.

- If the histogram has very narrow ranges for its groups, it looks like a big series of tiny bars that cloud the big picture. This can make the data look very choppy with no real pattern.

✔ Make sure your groups have equal widths. If one bar is wider than the others, it may contain more data than it should.

One idea that may be appropriate for your histogram is to take the range of the data (largest minus smallest) and divide by 10 to get 10 groupings.

Handling borderline values

In the Academy Award example, what happens if an actress's age lies right on a borderline? For example, in Table 7-1 Norma Shearer was 30 years old in 1930 when she won the Oscar for *The Divorcee*. Does she belong in the 25–30 age group (the lower bar) or the 30–35 age group (the upper bar)?

As long as you are consistent with all the data points, you can either put all the borderline points into their respective lower bars or put all of them into their respective upper bars. The important thing is to pick a direction and be consistent. In Figure 7-1, I went with the convention of putting all borderline values into their respective upper bars — which puts Norma Shearer's age in the 3rd bar, the 30–35 age group of Figure 7-1.

Clarifying the axes

The most complex part of interpreting a histogram for the reader is to get a handle on what's being shown on the *x* and *y* axes. Having good descriptive labels on the axes will help. Most statistical software packages label the *x*-axis using the variable name you provided when you entered your data (for example "age" or "weight"). However, the label for the *y*-axis isn't as clear. Statistical software packages often label the *y*-axis of a histogram by writing "frequency" or "percent" by default. These terms can be confusing: frequency or percentage of what?

Clarify the *y*-axis label on your histogram by changing "frequency" to "number of" and adding the variable name. To modify a label that simply reads "percent," clarify by writing "percentage of" and the variable. For example, in the histogram of ages of the Best Actress winners shown in Figure 7-1, I labeled the *y*-axis "Percentage of actresses in each age group." In the next section you see how to interpret the results from a histogram. How old are those actresses anyway?

Interpreting a histogram

A histogram tells you three main features of numerical data:

✔ How the data are distributed among the groups (statisticians call this the *shape* of the data)

✔ The amount of variability in the data (statisticians call this the amount of *spread* in the data)

✔ Where the center of the data is (statisticians use different measures)

Checking out the shape of the data

One of the features that a histogram can show you is the *shape* of the data — in other words, the manner in which the data fall into the groups. For example, all the data may be exactly the same, in which case the histogram is just one tall bar; or the data might have an equal number in each group; in which case the shape is flat.

Some data sets have a distinct shape. Here are three shapes that stand out:

✔ **Symmetric:** A histogram is symmetric if you cut it down the middle and the left-hand and right-hand sides resemble mirror images of each other.

Figure 7-2a shows a symmetric data set; it represents the amount of time each of 50 survey participants took to fill out a certain survey. You see that the histogram is close to symmetric.

✔ **Skewed right:** A skewed right histogram looks like a lopsided mound, with a tail going off to the right.

Figure 7-1, showing the ages of the Best Actress Award winners, is skewed right. You see on the right side there are a few actresses whose ages are older than the rest.

✔ **Skewed left:** If a histogram is skewed left, it looks like a lopsided mound with a tail going off to the left.

Figure 7-2b shows a histogram of 17 exam scores. The shape is skewed left; you see a few students who scored lower than everyone else.

Following are some particulars about classifying the shape of a data set:

✔ **Don't expect symmetric data to have an exact and perfect shape.** Data hardly ever fall into perfect patterns, so you have to decide whether the data shape is close enough to be called symmetric.

If the shape is close enough to symmetric that another person would notice it, and the differences aren't enough to write home about, I'd classify it as symmetric or roughly symmetric. Otherwise, you classify the data as non-symmetric. (More sophisticated statistical procedures exist that actually test data for symmetry, but they're beyond the scope of this book.)

✔ **Don't assume that data are skewed if the shape is non-symmetric.** Data sets come in all shapes and sizes, and many of them don't have a distinct shape at all. I include skewness on the list here because it's one of the more common non-symmetric shapes, and it's one of the shapes included in a standard introductory statistics course.

If a data set does turn out to be skewed (or close to it), make sure to denote the direction of the skewness (left or right).

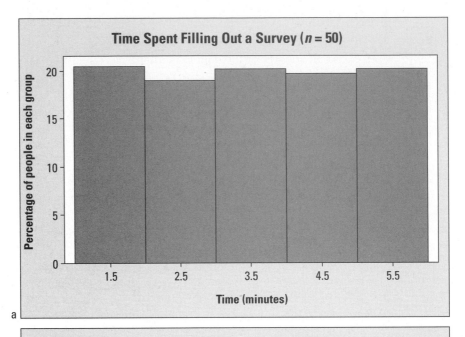

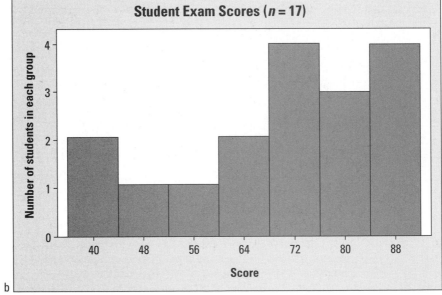

Figure 7-2:
Comparing the shape of a) a symmetric histogram and b) a skewed left histogram.

As you know from Figure 7-1, the actresses' ages in Figure 7-1 are skewed right. Most of the actresses were between 20 and 50 years of age when they won, with about 27% of them between the ages of 30–35. A few actresses were older when they won their Oscars; about 6 percent were between 60–65 years of age, and less than 4% (total) were 70 years old or over (if you add the percentages from the last two bars in the histogram). The last three bars are what make the data have a shape that is skewed right.

Measuring center: Mean versus median

A histogram gives you a rough idea of where the "center" of the data lies. The word *center* is in quotes because many different statistics are used to designate center. The two most common measures of center are the average (the mean) and the median. (For details on measures of center, see Chapter 5.)

To visualize the average age (the mean), picture the data as people sitting on a teeter-totter. Your objective is to balance it. Because data don't move around, assume the people stay where they are and you move the pivot point (which you can also think of as the hinge or fulcrum) anywhere you want. The mean is the place the pivot point has to be in order to balance the weight on each side of the teeter-totter.

The balancing point of the teeter-totter is affected by the weights of the people on each side, not by the number of people on each side. So the mean is affected by the actual values of the data, rather than the amount of data.

The median is the place where you put the pivot point so you have an equal number of people on each side of the teeter-totter, regardless of their weights. With the same number of people on each side, the teeter-totter wouldn't balance in terms of weight unless the teeter-totter had people with the same total weight on each side. So the median isn't affected by the values of the data, just their location within the data set.

The mean is affected by *outliers,* values in the data set that are away from the rest of the data, on the high end and/or the low end. The median, being the middle number, is not affected by outliers.

Viewing variability: Amount of spread around the mean

You also get a sense of variability in the data by looking at a histogram. For example, if the data are all the same, they are all placed into a single bar, and there is no variability. If an equal amount of data is in each group, the histogram looks flat with the bars close to the same height; this means a fair amount of variability.

The idea of a flat histogram indicating some variability may go against your intuition, and if it does you're not alone. If you're thinking a flat histogram means no variability, you're probably thinking about a time chart, where single numbers are plotted over time (see the section "Tackling Time Charts" later in this chapter). Remember, though, that a histogram doesn't show data over time — it shows all the data at one point in time.

Equally confusing is the idea that a histogram with a big lump in the middle and tails sloping sharply down on each side actually has less variability than a histogram that's straight across. The curves looking like hills in a histogram represent clumps of data that are close together; a flat histogram shows data equally dispersed, with more variability.

Variability in a histogram is higher when the taller bars are more spread out around the mean and lower when the taller bars are close to the mean.

For the Best Actress Award winners' ages shown in Figure 7-1, you see many actresses are in the age range from 30–35, and most of the ages are between 20–50 years in age, which is quite diverse; then you have those outliers, those few older actresses (I count 7 of them) that spread the data out farther, increasing its overall variability.

The most common statistic used to measure variability in a data set is the *standard deviation,* which in a rough sense measures the average distance that the data lie from the mean. The standard deviation for the Best Actress age data is 11.35 years. (See Chapter 5 for all the details on standard deviation.) A standard deviation of 11.35 years is fairly large in the context of this problem, but the standard deviation is based on average distance from the mean, and the mean is influenced by outliers, so the standard deviation will be as well (see Chapter 5 for more information).

In the later section "Interpreting a boxplot," I discuss another measure of variability, called the *interquartile range (IQR),* which is a more appropriate measure of variability when you have skewed data.

Putting numbers with pictures

You can't actually calculate measures of center and variability from the histogram itself because you don't know the exact data values. To add detail to your findings, you should always calculate the basic statistics of center and variation along with your histogram. (All the descriptive statistics you need, and then some, appear in Chapter 5.)

Figure 7-1 is a histogram for the Best Actress ages; you can see it is skewed right. Then for Figure 7-3, I calculated some basic (that is, descriptive) statistics from the data set. Examining these numbers, you find the median age is 33.00 years and the mean age is 35.69 years.

The mean age is higher than the median age because of a few actresses that were quite a bit older than the rest when they won their awards. For example, Jessica Tandy won for her role in *Driving Miss Daisy* when she was 81, and Katharine Hepburn won the Oscar for *On Golden Pond* when she was 74. The relationship between the median and mean confirms the skewness (to the right) found in Figure 7-1.

Figure 7-3:
Descriptive
statistics
for Best
Actress
ages
(1928–2009).

Descriptive Statistics: Age

Variable	Total Count	Mean	StDev	Minimum	Q1	Median	Q3	Maximum	IQR
Age	83	35.69	11.35	21.00	28.00	33.00	39.00	81.00	11.00

Here are some tips for connecting the shape of the histogram (discussed in the previous section) with the mean and median:

✔ **If the histogram is skewed right, the mean is greater than the median.**

This is the case because skewed-right data have a few large values that drive the mean upward but do not affect where the exact middle of the data is (that is, the median). Looking at the histogram of ages of the Best Actress Award winners in Figure 7-1, you see they're skewed right.

✔ **If the histogram is close to symmetric, then the mean and median are close to each other.**

Close to symmetric means it's almost the same on either side; it doesn't need to be exact. *Close* is defined in the context of the data; for example, the numbers 50 and 55 are said to be close if all the values lie between 0 and 1,000, but they are considered to be farther apart if all the values lie between 49 and 56.

The histogram shown in Figure 7-2a is close to symmetric. Its mean and median are both equal to 3.5.

✔ **If the histogram is skewed left, the mean is less than the median.**

This is the case because skewed-left data have a few small values that drive the mean downward but do not affect where the exact middle of the data is (that is, the median).

Figure 7-2b represents the exam scores of 17 students, and the data are skewed left. I calculated the mean and median of the original data set to be 70.41 and 74.00, respectively. The mean is lower than the median due to a few students who scored quite a bit lower than the others. These findings match the general shape of the histogram shown in Figure 7-2b.

The tips for interpreting histograms found in the previous section can also be used the other way around. If for some reason you don't have a histogram of the data, and you only have the mean and median to go by, you compare them to each other to get a rough idea as to the shape of the data set.

- If the mean is much larger than the median, the data are generally skewed right; a few values are larger than the rest.

- If the mean is much smaller than the median, the data are generally skewed left; a few smaller values bring the mean down.

- If the mean and median are close, you know the data is fairly balanced, or symmetric, on each side.

Under certain conditions, you can put together the mean and standard deviation to describe a data set in quite a bit of detail. If the data have a normal distribution (a bell-shaped hill in the middle, sloping down at the same rate on each side; see Chapter 5), the Empirical Rule can be applied.

The Empirical Rule (also in Chapter 5) says that if the data have a normal distribution, about 68% of the data lie within 1 standard deviation of the mean, about 95% of the data lie within 2 standard deviations from the mean, and 99.7% of the data lie within 3 standard deviations of the mean. These percentages are custom-made for the normal distribution (bell-shaped data) only and can't be used for data sets of other shapes.

Detecting misleading histograms

There are no hard and fast rules for how to create a histogram; the person making the graph gets to choose the groupings on the *x*-axis as well as the scale and starting and ending points on the *y*-axis. Just because there is an element of choice, however, doesn't mean every choice is appropriate; in fact, a histogram can be made to be misleading in many ways. In the following sections, you see examples of misleading histograms and how to spot them.

Missing the mark with too few groups

Although the number of groups you use for a histogram is up to the discretion of the person making the graph, there is such a thing as going overboard, either by having way too few bars, with everything lumped together, or by having way too many bars, where every little difference is magnified.

To decide how many bars a histogram should have, I take a good look at the groupings used to form the bars on the *x*-axis and see if they make sense. For example, it doesn't make sense to talk about exam scores in groups of 2 points; that's too much detail — too many bars. On the other hand, it doesn't make sense to group actresses' ages by intervals of 20 years; that's not descriptive enough.

Figures 7-4 and 7-5 illustrate this point. Each histogram summarizes *n* = 222 observations of the amount of time between eruptions of the Old Faithful geyser in Yellowstone Park. Figure 7-4 uses six bars that group the data by

10-minute intervals. This histogram shows a general skewed left pattern, but with 222 observations you are cramming an awful lot of data into only six groups; for example, the bar for 75–85 minutes has more than 90 pieces of data in it. You can break it down further than that.

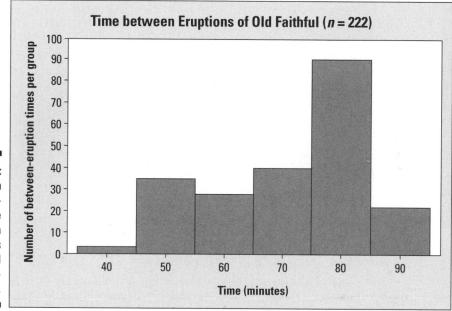

Figure 7-4:
Histogram #1 showing time between eruptions for Old Faithful geyser (*n* = 222).

Figure 7-5 is a histogram of the same data set, where the time between eruptions is broken into groups of 3 minutes each, resulting in 19 bars. Notice the distinct pattern in the data that shows up with this histogram which wasn't uncovered in Figure 7-4. You see two distinct peaks in the data; one peak around the 50-minute mark, and one around the 75-minute mark. A data set with two peaks is called *bimodal;* Figure 7-5 shows a clear example.

Looking at Figure 7-5, you can conclude that the geyser has two categories of eruptions; one group that has a shorter waiting time, and another group that has a longer waiting time. Within each group you see the data are fairly close to where the peak is located. Looking at Figure 7-4, you couldn't say that.

If the interval for the groupings of the numerical variable is really small, you see too many bars in the histogram; the data may be hard to interpret because the heights of the bars look more variable than they should be. On the other hand, if the ranges are really large, you see too few bars, and you may miss something interesting in the data.

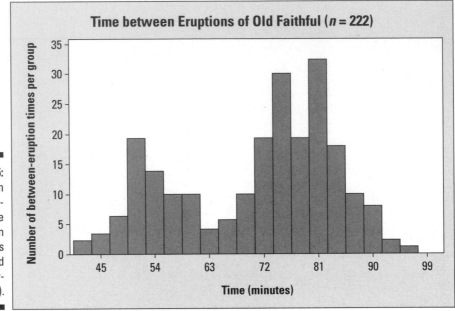

Time between Eruptions of Old Faithful (*n* = 222)

Figure 7-5:
Histogram
#2 show-
ing time
between
eruptions
for Old
Faithful gey-
ser (*n* = 222).

Watching the scale and start/finish lines

The *y*-axis of a histogram shows how many individuals are in each group, using counts or percents. A histogram can be misleading if it has a deceptive scale and/or inappropriate starting and ending points on the *y*-axis.

Watch the scale on the *y*-axis of a histogram. If it goes by large increments and has an ending point that's much higher than needed, you see a great deal of white space above the histogram. The heights of the bars are squeezed down, making their differences look more uniform than they should. If the scale goes by small increments and ends at the smallest value possible, the bars become stretched vertically, exaggerating the differences in their heights and suggesting a bigger difference than really exists.

An example comparing scales on the vertical (*y*) axes is shown in Figures 7-5 and 7-6. I took the Old Faithful data (time between eruptions) and made a histogram with vertical increments of 20 minutes, from 0 to 100; see Figure 7-6. Compare this to Figure 7-5, with vertical increments of 5 minutes, from 0 to 35. Figure 7-6 has a lot of white space and gives the appearance that the times are more evenly distributed among the groups than they really are. It also makes the data set look smaller, if you don't pay attention to what's on the *y*-axis. Of the two graphs, Figure 7-5 is more appropriate.

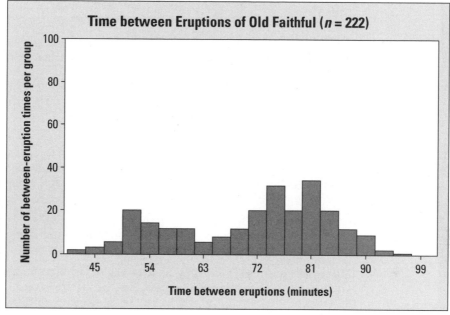

Figure 7-6:
Histogram
#3 of Old
Faithful
geyser
eruption
times.

Examining Boxplots

A *boxplot* is a one-dimensional graph of numerical data based on the five-number summary, which includes the minimum value, the 25th percentile (known as Q_1), the median, the 75th percentile (Q_3), and the maximum value. In essence, these five descriptive statistics divide the data set into four parts; each part contains 25% of the data. (See Chapter 5 for a full discussion of the five-number summary.)

Making a boxplot

To make a boxplot, follow these steps:

1. **Find the five-number summary of your data set. (Use the steps outlined in Chapter 5.)**

2. **Create a vertical (or horizontal) number line whose scale includes the numbers in the five-number summary and uses appropriate units of equal distance from each other.**

3. **Mark the location of each number in the five-number summary just above the number line (for a horizontal boxplot) or just to the right of the number line (for a vertical boxplot).**

4. **Draw a box around the marks for the 25th percentile and the 75th percentile.**

5. **Draw a line in the box where the median is located.**

6. **Determine whether or not outliers are present.**

   To make this determination, calculate the *IQR* (by subtracting $Q_3 - Q_1$); then multiply by 1.5. Add this amount to the value of Q_3 and subtract this amount from Q_1. This gives you a wider boundary around the median than the box does. Any data points that fall outside this boundary are determined to be outliers.

7. **If there are no outliers (according to your results of Step 6), draw lines from the upper and lower edges of the box out to the minimum and maximum values in the data set.**

8. **If there are outliers (according to your results of Step 6), indicate their location on the boxplot with * signs. Instead of drawing a line from the edge of the box all the way to the most extreme outlier, stop the line at the last data value that isn't an outlier.**

Many if not most software packages indicate outliers in a data set by using an asterisk (*) or star symbol and use the procedure outlined in Step 6 to identify outliers. However, not all packages use these symbols and procedures; check to see what your package does before analyzing your data with a boxplot.

A horizontal boxplot for ages of the Best Actress Oscar award winners from 1928–2009 is shown in Figure 7-7. You can see the numbers separating sections of the boxplot match the five-number summary statistics shown in Figure 7-3.

Boxplots can be vertical (straight up and down) with the values on the axis going from bottom (lowest) to top (highest); or they can be horizontal, with the values on the axis going from left (lowest) to right (highest). The next section shows you how to interpret a boxplot.

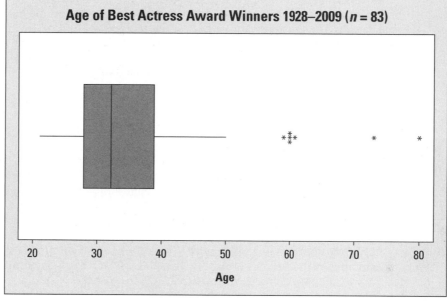

Figure 7-7: Boxplot of Best Actress ages (1928–2009; *n* = 83 actresses).

Interpreting a boxplot

Similar to a histogram (see the section "Interpreting a histogram"), a boxplot can give you information regarding the shape, center, and variability of a data set. Boxplots differ from histograms in terms of their strengths and weaknesses, as you see in the upcoming sections, but one of their biggest strengths is how they handle skewed data.

Checking the shape with caution!

A boxplot can show whether a data set is symmetric (roughly the same on each side when cut down the middle) or skewed (lopsided). A symmetric data set shows the median roughly in the middle of the box. Skewed data show a lopsided boxplot, where the median cuts the box into two unequal pieces. If the longer part of the box is to the right (or above) the median, the data is said to be *skewed right*. If the longer part is to the left (or below) the median, the data is *skewed left*.

As shown in the boxplot of the data in Figure 7-7, the ages are skewed right. The part of the box to the left of the median (representing the younger actresses) is shorter than the part of the box to the right of the median (representing the older actresses). That means the ages of the younger actresses are closer together than the ages of the older actresses. Figure 7-3 shows the descriptive statistics of the data and confirms the right skewness: the median age (33 years) is lower than the mean age (35.69 years).

If one side of the box is longer than the other, it does not mean that side contains more data. In fact, you can't tell the sample size by looking at a boxplot; it's based on percentages, not counts. Each section of the boxplot (the minimum to Q_1, Q_1 to the median, the median to Q_3, and Q_3 to the maximum) contains 25% of the data no matter what. If one of the sections is longer than another, it indicates a wider range in the values of data in that section (meaning the data are more spread out). A smaller section of the boxplot indicates the data are more condensed (closer together).

Although a boxplot can tell you whether a data set is symmetric (when the median is in the center of the box), it can't tell you the shape of the symmetry the way a histogram can. For example, Figure 7-8 shows histograms from two different data sets, each one containing 18 values that vary from 1 to 6. The histogram on the left has an equal number of values in each group, and the one on the right has two peaks at 2 and 5. Both histograms show the data are symmetric, but their shapes are clearly different.

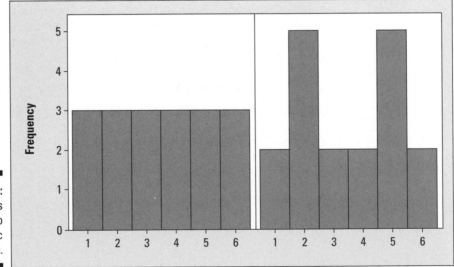

Figure 7-8:
Histograms
of two
symmetric
data sets.

Figure 7-9 shows the corresponding boxplots for these same two data sets; notice they are exactly the same. This is because the data sets both have the same five-number summaries — they're both symmetric with the same amount of distance between Q_1, the median, and Q_3. However, if you just saw the boxplots and not the histograms, you might think the shapes of the two data sets are the same, when indeed they are not.

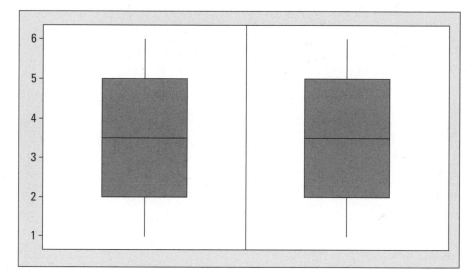

Figure 7-9:
Boxplots of the two symmetric data sets from Figure 7-8.

Despite its weakness in detecting the type of symmetry (you can add in a histogram to your analyses to help fill in that gap), a boxplot has a great upside in that you can identify actual measures of spread and center directly from the boxplot, where on a histogram you can't. A boxplot is also good for comparing data sets by showing them on the same graph, side by side.

All graphs have strengths and weaknesses; it's always a good idea to show more than one graph of your data for that reason.

Measuring variability with IQR

Variability in a data set that is described by the five-number summary is measured by the interquartile range (*IQR*). The *IQR* is equal to $Q_3 - Q_1$, the difference between the 75th percentile and the 25th percentile (the distance covering the middle 50% of the data). The larger the *IQR*, the more variable the data set is.

From Figure 7-3, the variability in age of the Best Actress winners as measured by the *IQR* is $Q_3 - Q_1 = 39 - 28 = 11$ years. Of the group of actresses whose ages were closest to the median, half of them were within 11 years of each other when they won their awards.

Notice that the *IQR* ignores data below the 25th percentile or above the 75th, which may contain outliers that could inflate the measure of variability of the entire data set. So if data is skewed, the *IQR* is a more appropriate measure of variability than the standard deviation.

Picking out the center using the median

The median, part of the five-number summary, is shown by the line that cuts through the box in the boxplot. This makes it very easy to identify. The mean, however, is not part of the boxplot and can't be determined accurately by just looking at the boxplot.

You don't see the mean on a boxplot because boxplots are based completely on percentiles. If data are skewed, the median is the most appropriate measure of center. Of course you can calculate the mean separately and add it to your results; it's never a bad idea to show both.

Investigating Old Faithful's boxplot

The relevant descriptive statistics for the Old Faithful geyser data are found in Figure 7-10.

Figure 7-10: Descriptive statistics for Old Faithful data.

Descriptive Statistics: Time between Eruptions

Variable	Total Count	Mean	StDev	Minimum	Q1	Median	Q3	Maximum	IQR
Time between	222	71.009	12.799	42.000	60.000	75.000	81.000	95.000	21.000

You can predict from the data set that the shape will be skewed left a bit because the mean is lower than the median by about 4 minutes. The *IQR* is $Q_3 - Q_1 = 81 - 60 = 21$ minutes, which shows the amount of overall variability in the time between eruptions; 50% of the eruptions are within 21 minutes of each other.

A vertical boxplot for length of time between eruptions of the Old Faithful geyser is shown in Figure 7-11. You confirm that the data are skewed left because the lower part of the box (where the small values are) is longer than the upper part of the box.

You see the values of the boxplot in Figure 7-11 that mark the five-number summary and the information shown in Figure 7-10, including the *IQR* of 21 minutes to measure variability. The center as marked by the median is 75 minutes; this is a better measure of center than the mean (71 minutes), which is driven down a bit by the left skewed values (the few that are shorter times than the rest of the data).

Looking at the boxplot (Figure 7-11), you see there are no outliers denoted by stars. However, note that the boxplot doesn't pick up on the bimodal shape of the data that you see in Figure 7-5. You need a good histogram for that.

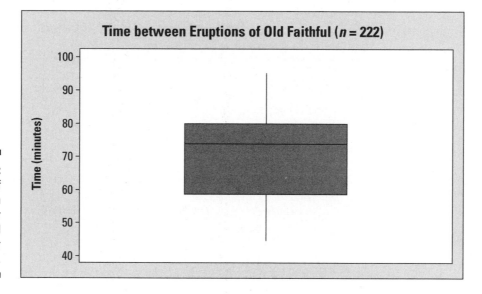

Figure 7-11:
Boxplot of
eruption
times for
Old Faithful
geyser
($n = 222$).

Denoting outliers

Looking at the boxplot in Figure 7-7 for the Best Actress ages data, you see a set of outliers (seven in all) on the right side of the data set, marked by a group of stars (as described in Step 8 in the earlier section "Making a boxplot"). Three of the stars lie on top of one another because three actresses were the same age, 61, when they won their Oscars.

You verify these outliers by applying the rule described in Step 6 of the section "Making a Boxplot." The *IQR* is 11 (from Figure 7-3), so you take $11 * 1.5 = 16.5$ years. Add this amount to Q_3 and you get $39 + 16.5 = 55.5$ years; subtracting this amount from Q_1 you get $28 - 16.5 = 11.5$ years. So an actress whose age was below 11.5 years (that is, 11 years old and under) or above 55.5 years (that is, 56 years old or over) is considered to be an outlier.

Of course, the lower end of this boundary (11.5 years) isn't relevant because the youngest actress was 21 (Figure 7-3 shows the minimum is 21). So you know there aren't any outliers on the low end of this data set.

However, seven outliers are on the high end of the data set, where the 56-and-over actresses' ages are. Table 7-2 shows the information on all seven outliers in the Best Actress ages data set.

Table 7-2		Best Actress Winners with Ages Designated as Outliers	
Year	Name	Age	Movie
1967	Katharine Hepburn	60	*Guess Who's Coming to Dinner*
1968	Katharine Hepburn	61	*The Lion in Winter*
1985	Geraldine Page	61	*Trip to Bountiful*
2006	Helen Mirren	61	*The Queen*
1931	Marie Dressler	62	*Min and Bill*
1981	Katharine Hepburn	74	*On Golden Pond*
1989	Jessica Tandy	81	*Driving Miss Daisy*

The youngest of the outliers is 60 years old (Katharine Hepburn, 1967). Just to compare, the next youngest age in the data set is 49 (Susan Sarandon, 1995). This indicates a clear break in this data set.

Making mistakes when interpreting a boxplot

It's a common mistake to associate the size of the box in a boxplot with the amount of data in the data set. Remember that each of the four sections shown in the boxplot contains an equal percentage (25%) of the data; the boxplot just marks off the places in the data set that separate those sections.

In particular, if the median splits the box into two unequal parts, the larger part contains data that's more variable than the other part, in terms of its range of values. However, there is still the same amount of data (25%) in the larger part of the box as there is in the smaller part.

Another common error involves sample size. A boxplot is a one-dimensional graph with only one axis representing the variable being measured. There is no second axis that tells you how many data points are in each group. So if you see two boxplots side-by-side and one of them has a very long box and the other has a very short one, don't conclude that the longer one has more data in it. The length of the box represents the variability in the data, not the number of data values.

When viewing or making a boxplot, always make sure the sample size (*n*) is included as part of the title. You can't figure out the sample size otherwise.

Tackling Time Charts

A *time chart* (also called a *line graph*) is a data display used to examine trends in data over time (also known as time series data). Time charts show time on the *x*-axis (for example, by month, year, or day) and the values of the variable being measured on the *y*-axis (like birth rates, total sales, or population size). Each point on the time chart summarizes all the data collected at that particular time; for example, the average of all pepper prices for January or the total revenue for 2010.

Interpreting time charts

To interpret a time chart, look for patterns and trends as you move across the chart from left to right.

The time chart in Figure 7-12 shows the ages of the Best Actress winners, in order of year won, from 1928–2009. Each dot indicates the age of a single actress, the one that won the Oscar that year. You see a bit of a cyclical pattern across time; that is, the ages go up, down, up, down, up, down with at least some regularity. It's hard to say what may be going on here; many variables go into determining an Oscar winner, including the type of movie, type of female role, mood of the voters, and so forth, and some of these variables may have a cyclical pattern to them.

Figure 7-12 also shows a very faint trend in age that is tending uphill; indicating that the Best Actress Award winners may be winning their awards increasingly later in life. Again, I wouldn't make too many assumptions from this result because the data has a great deal of variability.

As far as variability goes, you see that the ages represented by the dots do fluctuate quite a bit on the *y*-axis (representing age); all the dots basically fall between 20 and 80 years, with most of them between 25 and 45 years, I'd say. This goes along with the descriptive statistics found in Figure 7-3.

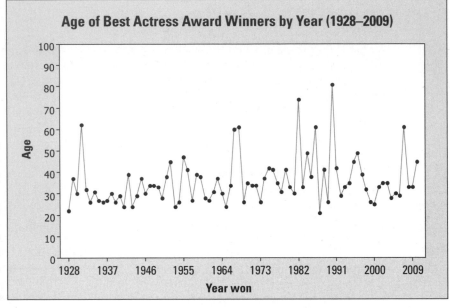

Figure 7-12:
Time Chart
#1 for ages
of Best
Actress
Academy
Award
winners,
1928–2009.

Understanding variability: Time charts versus histograms

Variability in a histogram should not be confused with variability in a time chart. If values change over time, they're shown on a time chart as highs and lows, and many changes from high to low (over time) indicate lots of variability. So a flat line on a time chart indicates no change and no variability in the values across time. For example, if the price of a product stays the same for 12 months in a row, the time chart for price would be flat.

But when the heights of a histogram's bars appear flat, the data is spread out uniformly across all the groups, indicating a great deal of variability in the data. (For an example, refer to Figure 7-2a.)

Spotting misleading time charts

As with any graph, you have to evaluate the units of the numbers being plotted. For example, it's misleading to chart the *number* of crimes over time, rather than the crime *rate* (crimes per capita) — because the population size of a city changes over time, crime rate is the appropriate measure. Make sure you understand what numbers are being graphed and examine them for fairness and appropriateness.

Watching the scale and start/end points

The scale on the vertical axis can make a big difference in the way the time chart looks. Refer to Figure 7-12 to see my original time chart of the ages for the Best Actress Academy Award winners from 1928–2009 in increments of 10 years. You see a fair amount of variability, as discussed previously.

In Figure 7-12, the starting and ending points on the vertical axis are 0 to 100, which creates a little bit of extra white space on the top and bottom of the picture. I could have used 10 and 90 as my start/end points, but this graph looks reasonable.

Now what happens if I change the vertical axis? Figure 7-13 shows the same data, with start/end points of 20 and 80. The increments of 10 years appear longer than the increments of 10 years shown in Figure 7-12. Both of these changes in the graph exaggerate the differences in ages even more.

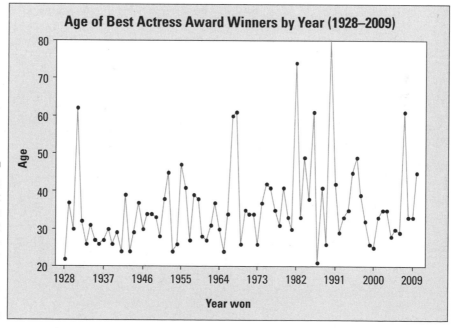

Figure 7-13:
Time Chart #2 for ages of Best Actress Oscar Award winners, 1928–2009.

How do you decide which graph is the best one for your data? There is no perfect graph; there is no right or wrong answer; but there are limits. You can quickly spot problems just by zooming in on the scale and start/end points.

Simplifying excess data

A time chart of the time between eruptions for the Old Faithful data is shown in Figure 7-14. You see 222 dots on this graph; each one represents the time between one eruption and the next, for every eruption during a 16-day period.

This figure looks very complex; data are everywhere, there are too many points to really see anything, and you can't find the forest for the trees. There is such a thing as having too much data, especially nowadays when you can measure data continuously and meticulously using all kinds of advanced technology. I'm betting they didn't have a student standing by the geyser recording eruption times on a clipboard, for example!

To get a clearer picture of the Old Faithful data, I combined all the observations from a single day and found its mean; I did this for all 16 days, and then I plotted all the means on a time chart in order. This reduced the data from 222 points to 16 points. The time chart is shown in Figure 7-15.

From this time chart I see a little bit of a cyclical pattern to the data; every day or two it appears to shift from short times between eruptions to longer times between eruptions. While these changes are not definitive, it does provide important information for scientists to follow up on when studying the behavior of geysers like Old Faithful.

Figure 7-14:
Time chart showing time between eruptions for Old Faithful Geyser ($n = 222$ consecutive observations).

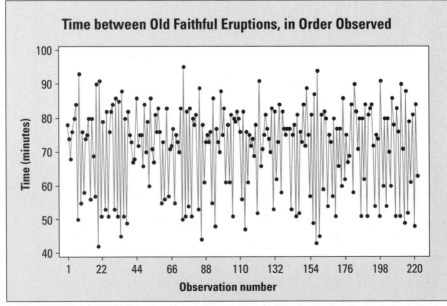

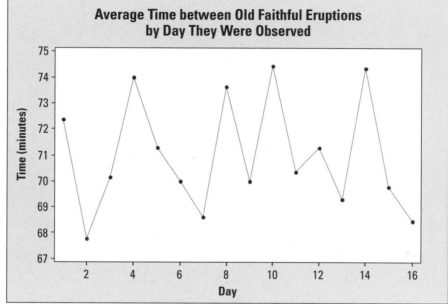

Figure 7-15:
Time chart showing daily average time between eruptions for Old Faithful geyser ($n = 16$ consecutive days).

A time chart condenses all the data for one unit of time into a single point. By contrast, a histogram displays the entire sample of data that was collected at that one unit of time. For example, Figure 7-15 shows the daily average time between eruptions for 16 days. For any given day, you can make a histogram of all the eruptions observed on that particular day. Displaying a time chart of average times over 16 days accompanied by a histogram summarizing all the eruptions for a particular day would be a great one-two punch.

Evaluating time charts

Here is a checklist for evaluating time charts, with a couple more thoughts added in:

✔ Examine the scale and start/end points on the vertical axis (the one showing the values of the data). Large increments and/or lots of white space make differences look less dramatic; small increments and/or a plot that totally fills the page exaggerate differences.

✔ If the amount of data you have is overwhelming, consider boiling it down by finding means/medians for blocks of time and plotting those instead.

✔ Watch for gaps in the timeline on a time chart. For example, it's misleading to show equally spaced points on the horizontal (time) axis for 1990, 2000, 2005, and 2010. This happens when years are just treated like labels, rather than real numbers.

✔ As with any graph, take the units into account; be sure they're appropriate for comparison over time. For example, are dollar amounts adjusted for inflation? Are you looking at number of crimes, or the crime rate?

Part III
Distributions and the Central Limit Theorem

The 5th Wave   By Rich Tennant

"This guy writes an equation for over 20 minutes,
and he has the nerve to say, 'Voilà'?"

In this part . . .

Statisticians study populations; that's their bread and butter. They measure, count, or classify characteristics of a population (using random variables); find probabilities and proportions; and create (or estimate) numerical summaries for the population (that is, parameters for the population). Sometimes you know a great deal about a population from the start; sometimes it's hazier. This part studies populations under both scenarios.

If a population fits a specific distribution, tools are available for studying it. In Chapters 8 through 10, you see three commonly used distributions: the binomial distribution (for categorical data) and the normal and *t*-distributions (for numerical data).

If the specifics about a population are unknown (as happens most of the time), you take a sample and generalize its results to the population. However, sample results vary, and you need to take that into account. In Chapter 11 you investigate sample variability, measure the precision of your sample results, and find probabilities for their likelihood. From there you'll be able to properly estimate parameters and test claims made about them, but that's another Part — IV, to be exact.

Chapter 8

Random Variables and the Binomial Distribution

Scientists and engineers often build models for the phenomena they are studying to make predictions and decisions. For example, where and when is this hurricane going to hit when it makes landfall? How many accidents will occur at this intersection this year if it's not redone? Or, what will the deer population be like in a certain region five years from now?

To answer these questions, scientists (usually working with statisticians) define a characteristic they are measuring or counting (such as number of intersections, location and time when a hurricane hits, population size, and so on) and treat it as a variable that changes in some random way, according to a certain pattern. They cleverly call them — you guessed it — random variables. In this chapter, you find out more about random variables, their types and characteristics, and why they are important. And you look at the details of one of the most common random variables: the binomial.

Defining a Random Variable

A *random variable* is a characteristic, measurement, or count that changes randomly according to a certain set or pattern. Its notation is X, Y, Z, and so on. In this section, you see how different random variables are characterized and how they behave in the long term in terms of their means and standard deviations.

In math you have variables like *X* and *Y* that take on certain values depending on the problem (for example, the width of a rectangle), but in statistics the variables change in a random way. By *random,* statisticians mean that you don't know exactly what the next outcome will be but you do know that certain outcomes happen more frequently than others; everything's not 50-50. (Like when I try to shoot baskets; it's definitely not a 50% chance I'll make one and 50% chance I'll miss. It's more like 5% chance of making it and a 95% chance of missing it.) You can use that information to better study data and populations and make good decisions. (For example, don't put me in your basketball game to shoot free throws.)

Data have different types: categorical and numerical (see Chapter 4). While both types of data are associated with random variables, I discuss only numerical random variables here (this falls in line with most intro stat courses as well). For information on analyzing categorical variables, see Chapters 6 and 19.

Discrete versus continuous

Numerical random variables represent counts and measurements. They come in two different flavors: discrete and continuous, depending on the type of outcomes that are possible.

- ✔ **Discrete random variables:** If the possible outcomes of a random variable can be listed out using whole numbers (for example, 0, 1, 2 . . . , 10; or 0, 1, 2, 3), the random variable is *discrete.*

- ✔ **Continuous random variables:** If the possible outcomes of a random variable can only be described using an interval of real numbers (for example, all real numbers from zero to infinity), the random variable is *continuous.*

Discrete random variables typically represent counts — for example, the number of people who voted yes for a smoking ban out of a random sample of 100 people (possible values are 0, 1, 2, . . . , 100); or the number of accidents at a certain intersection over one year's time (possible values are 0, 1, 2, . . .).

Discrete random variables have two classes: finite and countably infinite. A discrete random variable is *finite* if its list of possible values has a fixed (finite) number of elements in it (for example, the number of smoking ban supporters in a random sample of 100 voters has to be between 0 and 100). One very common finite random variable is the binomial, which is discussed in this chapter in detail.

A discrete random variable is *countably infinite* if its possible values can be specifically listed out but they have no specific end. For example, the number of accidents occurring at a certain intersection over a 10-year period can take on possible values: 0, 1, 2, . . . (you know they end somewhere but you can't say where, so you list them all).

Continuous random variables typically represent measurements, such as time to complete a task (for example 1 minute 10 seconds, 1 minute 20 seconds, and so on) or the weight of a newborn. What separates continuous random variables from discrete ones is that they are *uncountably infinite;* they have too many possible values to list out or to count and/or they can be measured to a high level of precision (such as the level of smog in the air in Los Angeles on a given day, measured in parts per million).

Examples of commonly used continuous random variables can be found in Chapter 9 (the normal distribution) and Chapter 10 (the *t*-distribution).

Probability distributions

A discrete random variable X can take on a certain set of possible outcomes, and each of those outcomes has a certain probability of occurring. The notation used for any specific outcome is a lowercase x. For example, say you roll a die and look at the outcome. The random variable X is the outcome of the die (which takes on possible values of 1, 2, . . . , 6). Now if you roll the die and get a 1, that's a specific outcome, so you write "$x = 1$."

The probability of any specific outcome occurring is denoted $p(x)$, which you pronounce "*p* of *x*." It signifies the probability that the random variable X takes on a specific value, which you call "little *x*." For example, to denote the probability of getting a 1 on a die, you write $p(1)$.

Statisticians use an uppercase X when they talk about random variables in their general form; for example, "Let X be the outcome of the roll of a single die." They use lowercase x when they talk about specific outcomes of the random variable, like $x = 1$ or $x = 2$.

A list or function showing all possible values of a discrete random variable, along with their probabilities, is called a *probability distribution*, $p(x)$. For example, when you roll a single die, the possible outcomes are 1, 2, 3, 4, 5, and 6, and each has a probability of ⅙ (if the die is fair). As another example, suppose 40% of renters living in an apartment complex own one dog, 7% own two dogs, 3% own three dogs, and 50% own zero dogs. For X = the number of dogs owned, the probability distribution for X is shown in Table 8-1.

Table 8-1	Probability Distribution for X = Number of Dogs Owned by Apartment Renters
x	p(x)
0	0.50
1	0.40
2	0.07
3	0.03

The mean and variance of a discrete random variable

The *mean* of a random variable is the average of all the outcomes you would expect in the long term (over all possible samples). For example, if you roll a die a billion times and record the outcomes, the average of those outcomes is 3.5. (Each outcome happens with equal chance, so you average the numbers 1 through 6 to get 3.5.) However, if the die is loaded and you roll a 1 more often than anything else, the average outcome from a billion rolls is closer to 1 than to 3.5.

The notation for the mean of a random variable X is μ_x or μ (pronounced "mu sub x"; or just "mu x"). Because you are looking at all the outcomes in the long term, it's the same as looking at the mean of an entire population of values, which is why you denote it μ_x and not $\bar{x}$. (The latter represents the mean of a *sample* of values [see Chapter 5].) You put the X in the subscript to remind you that the variable this mean belongs to is the X variable (as opposed to a Y variable or some other letter).

The *variance* of a random variable is roughly interpreted as the average squared distance from the mean for all the outcomes you would get in the long term, over all possible samples. This is the same as the variance of the population of all possible values. The notation for variance of a random variable X is σ_x^2 or σ^2. You say "sigma sub x, squared" or just "sigma squared."

The standard deviation of a random variable X is the square root of the variance, denoted by σ_x or σ (say "sigma x" or just "sigma"). It roughly represents the average distance from the mean.

Just like for the mean, you use the Greek notation to denote the variance and standard deviation of a random variable. The English notation s^2 and s represent the variance and standard deviation of a *sample* of individuals, not the entire population (see Chapter 5).

The variance is in square units, so it can't be easily interpreted. You use standard deviation for interpretation because it is in the original units of X. The standard deviation can be roughly interpreted as the average distance away from the mean.

Identifying a Binomial

The most well-known and loved discrete random variable is the binomial. *Binomial* means *two names* and is associated with situations involving two outcomes; for example yes/no, or success/failure (hitting a red light or not, developing a side effect or not). This section focuses on the binomial random variable — when you can use it, finding probabilities for it, and finding its mean and variance.

A random variable is binomial (that is, it has a binomial distribution) if the following four conditions are met:

1. There are a fixed number of trials (n).

2. Each trial has two possible outcomes: success or failure.

3. The probability of success (call it p) is the same for each trial.

4. The trials are independent, meaning the outcome of one trial doesn't influence that of any other.

Let X equal the total number of successes in n trials; if all four conditions are met, X has a binomial distribution with probability of success (on each trial) equal to p.

The lowercase p here stands for the probability of getting a success on one single (individual) trial. It's not the same as $p(x)$, which means the probability of getting x successes in n trials.

Checking binomial conditions step by step

You flip a fair coin 10 times and count the number of heads (X). Does X have a binomial distribution? You can check by reviewing your responses to the questions and statements in the list that follows:

1. **Are there a fixed number of trials?**

   You're flipping the coin 10 times, which is a fixed number. Condition 1 is met, and $n = 10$.

2. **Does each trial have only two possible outcomes — success or failure?**

   The outcome of each flip is either heads or tails, and you're interested in counting the number of heads. That means success = heads, and failure = tails. Condition 2 is met.

3. **Is the probability of success the same for each trial?**

   Because the coin is fair, the probability of success (getting a head) is $p = \frac{1}{2}$ for each trial. You also know that $1 - \frac{1}{2} = \frac{1}{2}$ is the probability of failure (getting a tail) on each trial. Condition 3 is met.

4. **Are the trials independent?**

   You assume the coin is being flipped the same way each time, which means the outcome of one flip doesn't affect the outcome of subsequent flips. Condition 4 is met.

Because the random variable X (the number of successes [heads] that occur in 10 trials [flips]) meets all four conditions, you conclude it has a binomial distribution with $n = 10$ and $p = \frac{1}{2}$.

But not every situation that appears binomial actually is. Read on to see some examples of what I mean.

No fixed number of trials

Suppose that you're going to flip a fair coin until you get four heads and you'll count how many flips it takes to get there; in this case X = number of flips. This certainly sounds like a binomial situation: Condition 2 is met because you have success (heads) and failure (tails) on each flip; condition 3 is met with the probability of success (heads) being the same (0.5) on each flip; and the flips are independent, so condition 4 is met.

However, notice that X isn't counting the number of heads, it counts the number of trials needed to get 4 heads. The number of successes (X) is fixed rather than the number of trials (n). Condition 1 is not met, so X does not have a binomial distribution in this case.

More than success or failure

Some situations involve more than two possible outcomes, yet they can appear to be binomial. For example, suppose you roll a fair die 10 times and let X be the outcome of each roll (1, 2, 3, . . . , 6). You have a series of $n = 10$ trials, they are independent, and the probability of each outcome is the same for each roll. However, on each roll you're recording the outcome on a six-sided die, a number from 1 to 6. This is not a success/failure situation, so condition 2 is not met.

However, depending on what you're recording, situations originally having more than two outcomes can fall under the binomial category. For example, if you roll a fair die 10 times and each time you record whether or not you get a 1, then condition 2 is met because your two outcomes of interest are getting a 1 ("success") and not getting a 1 ("failure"). In this case, p (the probability of success) = $\frac{1}{6}$, and $\frac{5}{6}$ is the probability of failure. So if X is counting the number of 1s you get in 10 rolls, X is a binomial random variable.

Trials are not independent

The independence condition is violated when the outcome of one trial affects another trial. Suppose you want to know opinions of adults in your city regarding a proposed casino. Instead of taking a random sample of, say, 100 people, to save time you select 50 married couples and ask each of them what their opinion is. In this case it's reasonable to say couples have a higher chance of agreeing on their opinions than individuals selected at random, so the independence condition 4 is not met.

Probability of success (p) changes

You have 10 people — 6 women and 4 men — and you want to form a committee of 2 people at random. Let X be the number of women on the committee of 2. The chance of selecting a woman at random on the first try is $\frac{6}{10}$. Because you can't select this same woman again, the chance of selecting another woman is now $\frac{5}{9}$. The value of p has changed, and condition 3 is not met.

If the population is very large (for example all U.S. adults), p still changes every time you choose someone, but the change is negligible, so you don't worry about it. You still say the trials are independent with the same probability of success, p. (Life is so much easier that way!)

Finding Binomial Probabilities Using a Formula

After you identify that X has a binomial distribution (the four conditions from the section "Checking binomial conditions step by step" are met), you'll likely want to find probabilities for X. The good news is that you don't have to find them from scratch; you get to use established formulas for finding binomial probabilities, using the values of n and p unique to each problem. Probabilities for a binomial random variable X can be found using the following formula for $p(x)$:

$$\binom{n}{x} p^x (1-p)^{n-x}$$

where

> ✔ n is the fixed number of trials.
>
> ✔ x is the specified number of successes.
>
> ✔ $n - x$ is the number of failures.
>
> ✔ p is the probability of success on any given trial.
>
> ✔ $1 - p$ is the probability of failure on any given trial. (***Note:*** Some textbooks use the letter q to denote the probability of failure rather than $1 - p$.)

These probabilities hold for any value of X between 0 (lowest number of possible successes in n trials) and n (highest number of possible successes).

The number of ways to rearrange x successes among n trials is called "n choose x," and the notation is $\binom{n}{x}$. It's important to note that this math expression is not a fraction; it's math shorthand to represent the number of ways to do these types of rearrangements.

In general, to calculate "n choose x," you use the following formula:

$$\binom{n}{x} = \frac{n!}{x!(n-x)!}$$

The notation $n!$ stands for *n-factorial,* the number of ways to rearrange n items. To calculate $n!$, you multiply $n(n-1)(n-2) \ldots (2)(1)$. For example 5! is $5(4)(3)(2)(1) = 120$; 2! is $2(1) = 2$; and 1! is 1. By convention, 0! equals 1.

Suppose you have to cross three traffic lights on your way to work. Let X be the number of red lights you hit out of the three. How many ways can you hit two red lights on your way to work? Well, you could hit a green one first, then the other two red; or you could hit the green one in the middle and have red ones for the first and third lights, or you could hit red first, then another red, then green. Letting G = green and R=red, you can write these three possibilities as: GRR, RGR, RRG. So you can hit two red lights on your way to work in three ways, right?

Check the math. In this example, a "trial" is a traffic light; and a "success" is a red light. (I know, that seems weird, but a success is whatever you are interested in counting, good or bad.) So you have $n = 3$ total traffic lights, and you're interested in the situation where you get $x = 2$ red ones. Using the fancy notation, $\binom{3}{2}$ means "3 choose 2" and stands for the number of ways to rearrange 2 successes in 3 trials.

To calculate "3 choose 2," you do the following:

$$\binom{3}{2} = \frac{3!}{2!(3-2)!} = \frac{3(2)(1)}{[(2)(1)](1)} = \frac{6}{2} = 3$$

This confirms the three possibilities listed for getting two red lights.

Now suppose the lights operate independently of each other and each one has a 30% chance of being red. Suppose you want to find the probability distribution for X. (That is, a list of all possible values of X — 0,1,2,3 — and their probabilities.)

Before you dive into the calculations, you first check the four conditions (from the section "Checking binomial conditions step by step") to see if you have a binomial situation here. You have $n = 3$ trials (traffic lights) — check. Each trial is success (red light) or failure (yellow or green light; in other words, "non-red" light) — check. The lights operate independently, so you have the independent trials taken care of, and because each light is red 30% of the time, you know $p = 0.30$ for each light. So X = number of red traffic lights has a binomial distribution. To fill in the nitty gritties for the formulas, $1 - p$ = probability of a non-red light = $1 - 0.30 = 0.70$; and the number of non-red lights is $3 - X$.

Using the formula for $p(x)$, you obtain the probabilities for $x = 0$, 1, 2, and 3 red lights:

$$p(0) = \binom{3}{0} 0.30^0 (1 - 0.30)^{3-0} =$$

$$\frac{3!}{0!(3-0)!}(0.30)^0 (0.70)^3 = 1(1)(0.343) = 0.343;$$

$$p(1) = \binom{3}{1} 0.30^1 (1 - 0.30)^{3-1} =$$

$$\frac{3!}{1!(3-1)!}(0.30)^1 (0.70)^2 = 3(0.30)(0.49) = 0.441;$$

$$p(2) = \binom{3}{2} 0.30^2 (1 - 0.30)^{3-2} =$$

$$\frac{3!}{2!(3-2)!}(0.30)^2 (0.70)^1 = 3(0.09)(0.70) = 0.189; \text{ and}$$

$$p(3) = \binom{3}{3} 0.30^3 (1 - 0.30)^{3-3} =$$

$$\frac{3!}{3!(3-3)!}(0.30)^3 (0.70)^0 = 1(0.027)(1) = 0.027.$$

The final probability distribution for X is shown in Table 8-2. Notice these probabilities all sum to 1 because every possible value of X is listed and accounted for.

Table 8-2	Probability Distribution for X = Number of Red Traffic Lights ($n = 3$, $p = 0.30$)
X	$p(x)$
0	0.343
1	0.441
2	0.189
3	0.027

Finding Probabilities Using the Binomial Table

The previous section deals with values of n that are pretty small, but you may wonder how you are going to handle the formula for calculating binomial probabilities when n gets large. No worries! A large range of binomial probabilities are provided in the binomial table in the appendix. Here's how to use it:

Within the binomial table you see several mini-tables; each one corresponds with a different n for a binomial (n = 1, 2, 3, ..., 15, and 20 are available). Each mini-table has rows and columns. Running down the side of any mini-table, you see all the possible values of X from 0 through n, each with its own row. The columns of the binomial table represent various values of p from 0.10 through 0.90.

Finding probabilities for specific values of X

To use the binomial table in the appendix to find probabilities for X = total number of successes in n trials where p is the probability of success on any individual trial, follow these steps:

1. **Find the mini-table associated with your particular value of n (the number of trials).**

2. **Find the column that represents your particular value of *p* (or the one closest to it, if appropriate).**

3. **Find the row that represents the number of successes (*x*) you are interested in.**

4. **Intersect the row and column from Steps 2 and 3.** This gives you the probability for *x* successes, written as *p*(*x*).

For the traffic light example from "Finding Binomial Probabilities Using a Formula," you can use the binomial table (Table A-3 in the appendix) to verify the results found by the binomial formula shown back in Table 8-2. Go to the mini-table where *n* = 3 and look in the column where *p* = 0.30. You see four probabilities listed for this mini-table: 0.343, 0.441, 0.189, and 0.027; these are the probabilities for *X* = 0, 1, 2, and 3 red lights, respectively, matching those from Table 8-2.

Finding probabilities for X greater-than, less-than, or between two values

The binomial table (Table A-3 in the appendix) shows probabilities for *X* being equal to any value from 0 to *n*, for a variety of *p*s. To find probabilities for *X* being less-than, greater-than, or between two values, just find the corresponding values in the table and add their probabilities. For the traffic light example, you count the number of times (*X*) that you hit a red light (out of 3 possible lights). Each light has a 0.30 chance of being red, so you have a binomial distribution with *n* = 3 and *p* = 0.30. If you want the probability that you hit more than one red light, you find *p*(*x* > 1) by adding *p*(2) + *p*(3) from Table A-3 to get 0.189 + 0.027 = 0.216.

The probability that you hit between 1 and 3 (inclusive) red lights is $p(1 \leq x \leq 3) = 0.441 + 0.189 + 0.027 = 0.657$.

You have to distinguish between a *greater-than* (>) and a *greater-than-or-equal-to* (≥) probability when working with discrete random variables. Repackaging the previous two examples, you see $p(x > 1) = 0.216$ but $p(x \geq 1) = 0.657$. This is a non-issue for continuous random variables (see Chapter 9).

Other phrases to remember: *at least* means that number or higher, and *at most* means that number or lower. For example, the probability that *X* is at least 2 is $p(x \geq 2)$; the probability that *X* is at most 2 is $p(x \leq 2)$.

Checking Out the Mean and Standard Deviation of the Binomial

Because the binomial distribution is so commonly used, statisticians went ahead and did all the grunt work to figure out nice, easy formulas for finding its mean, variance, and standard deviation. (That is, they've already applied the methods from the section "Defining a Random Variable" to the binomial distribution formulas, crunched everything out, and presented the results to us on a silver platter — don't you love it when that happens?) The following results are what came out of it.

If X has a binomial distribution with n trials and probability of success p on each trial, then:

1. The mean of X is $\mu = np$.

2. The variance of X is $\sigma^2 = np(1-p)$.

3. The standard deviation of X is $\sigma = \sqrt{np(1-p)}$.

For example, suppose you flip a fair coin 100 times and let X be the number of heads; then X has a binomial distribution with $n = 100$ and $p = 0.50$. Its mean is $\mu = np = 100(0.50) = 50$ heads (which makes sense, because heads and tails are 50-50). The variance of X is $\sigma^2 = np(1-p) = 100(0.50)(1-0.50) = 25$, which is in square units (so you can't interpret it); and the standard deviation is the square root of the variance, which is 5. That means when you flip a coin 100 times, and do that over and over, the average number of heads you'll get is 50, and you can expect that to vary by about 5 heads on average.

The formula for the mean of a binomial distribution has intuitive meaning. The p in the formula represents the probability of a success, yes, but it also represents the *proportion* of successes you can expect in n trials. Therefore, the total *number* of successes you can expect — that is, the mean of X — is $\mu = np$.

The formula for variance has intuitive meaning as well. The only variability in the outcomes of each trial is between success (with probability p) and failure (with probability $1 - p$). Over n trials, the variance of the number of successes/failures is measured by $\sigma^2 = np(1-p)$. The standard deviation is just the square root.

If the value of n is too large to use the binomial formula or the binomial table to calculate probabilities (see the earlier sections in this chapter), there's an alternative. It turns out that if n is large enough, you can use the normal distribution to get an approximate answer for a binomial probability. The mean and standard deviation of the binomial are involved in this process. All the details are in Chapter 9.

Chapter 9

The Normal Distribution

*I*n your statistical travels you'll come across two major types of random variables: discrete and continuous. *Discrete random variables* basically count things (number of heads on 10 coin flips, number of female Democrats in a sample, and so on). The most well-known discrete random variable is the binomial. (See Chapter 8 for more on discrete random variables and binomials). A *continuous random variable* is typically based on measurements; it either takes on an uncountably infinite number of values (values within an interval on the real line), or it has so many possible values that it may as well be deemed continuous (for example, time to complete a task, exam scores, and so on).

In this chapter, you understand and calculate probabilities for the most famous continuous random variable of all time — the normal distribution. You also find percentiles for the normal distribution, where you are given a probability as a percent and you have to find the value of X that's associated with it. And you can think how funny it would be to see a statistician wearing a T-shirt that said "I'd rather be normal."

Exploring the Basics of the Normal Distribution

A continuous random variable X has a normal distribution if its values fall into a smooth (continuous) curve with a bell-shaped pattern. Each normal distribution has its own mean, denoted by the Greek letter μ (say "mu"); and its own standard deviation, denoted by the Greek letter σ (say "sigma"). But no matter what their means and standard deviations are, all normal distributions have the same basic bell shape. Figure 9-1 shows some examples of normal distributions.

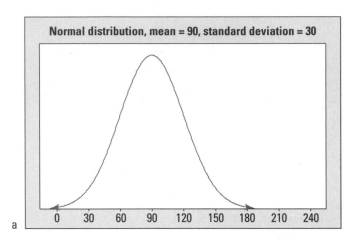

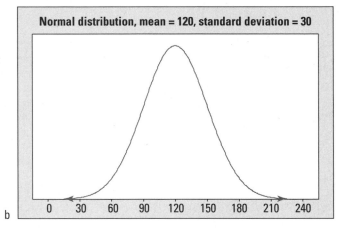

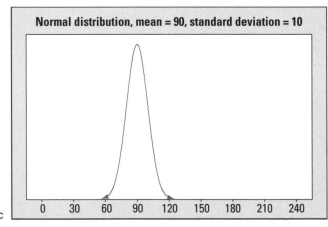

Figure 9-1:
Three normal distributions, with means and standard deviations of a) 90 and 30; b) 120 and 30; and c) 90 and 10, respectively.

Every normal distribution has certain properties. You use these properties to determine the relative standing of any particular result on the distribution, and to find probabilities. The properties of any normal distribution are as follows:

- ✔ Its shape is symmetric (that is, when you cut it in half the two pieces are mirror images of each other).

- ✔ Its distribution has a bump in the middle, with tails going down and out to the left and right.

- ✔ The mean and the median are the same and lie directly in the middle of the distribution (due to symmetry).

- ✔ Its standard deviation measures the distance on the distribution from the mean to the *inflection point* (the place where the curve changes from an "upside-down-bowl" shape to a "right-side-up-bowl" shape).

- ✔ Because of its unique bell shape, probabilities for the normal distribution follow the Empirical Rule (full details in Chapter 5), which says the following:

  - About 68 percent of its values lie within one standard deviation of the mean. To find this range, take the value of the standard deviation, then find the mean plus this amount, and the mean minus this amount.

  - About 95 percent of its values lie within two standard deviations of the mean. (Here you take 2 times the standard deviation, then add it to and subtract it from the mean.)

  - Almost all of its values (about 99.7 percent of them) lie within three standard deviations of the mean. (Take 3 times the standard deviation and add it to and subtract it from the mean.)

- ✔ Precise probabilities for all possible intervals of values on the normal distribution (not just for those within 1, 2, or 3 standard deviations from the mean) are found using a table with minimal (if any) calculations. (The next section gives you all the info on this table.)

Take a look again at Figure 9-1. To compare and contrast the distributions shown in Figure 9-1a, b, and c, you first see they are all symmetric with the signature bell shape. The examples in Figure 9-1a and Figure 9-1b have the same standard deviation, but their means are different; Figure 9-1b is located 30 units to the right of Figure 9-1a because its mean is 120 compared to 90. Figures 9-1a and c have the same mean (90), but Figure 9-1a has more variability than Figure 9-1c due to its higher standard deviation (30 compared to 10). Because of the increased variability, the values in Figure 9-1a stretch from 0 to 180 (approximately), while the values in Figure 9-1c only go from 60 to 120.

Finally, Figures 9-1b and c have different means and different standard deviations entirely; Figure 9-1b has a higher mean which shifts it to the right, and Figure 9-1c has a smaller standard deviation; its values are the most concentrated around the mean.

Noting the mean and standard deviation is important so you can properly interpret numbers located on a particular normal distribution. For example, you can compare where the number 120 falls on each of the normal distributions in Figure 9-1. In Figure 9-1a, the number 120 is one standard deviation above the mean (because the standard deviation is 30, you get 90 + 1 * 30 = 120). So on this first distribution, the number 120 is the upper value for the range where about 68% of the data are located, according to the Empirical Rule (see Chapter 5).

In Figure 9-1b, the number 120 lies directly on the mean, where the values are most concentrated. In Figure 9-1c, the number 120 is way out on the rightmost fringe, 3 standard deviations above the mean (because the standard deviation this time is 10, you get 90 + 3[10]=120). In Figure 9-1c, values beyond 120 are very unlikely to occur because they are beyond the range where about 99.7% of the values should be, according to the Empirical Rule.

Meeting the Standard Normal (Z-) Distribution

One very special member of the normal distribution family is called the standard normal distribution, or Z-distribution. The *Z-distribution* is used to help find probabilities and percentiles for regular normal distributions (*X*). It serves as the standard by which all other normal distributions are measured.

Checking out Z

The Z-distribution is a normal distribution with mean zero and standard deviation 1; its graph is shown in Figure 9-2. Almost all (about 99.7%) of its values lie between –3 and +3 according to the Empirical Rule. Values on the Z-distribution are called *z*-values, *z*-scores, or standard scores. A *z-value* represents the number of standard deviations that a particular value lies above or below the mean. For example, $z = 1$ on the Z-distribution represents a value that is 1 standard deviation above the mean. Similarly, $z = -1$ represents a value that is one standard deviation below the mean (indicated by the minus sign on the *z*-value). And a *z*-value of 0 is — you guessed it — right on the mean. All *z*-values are universally understood.

If you refer back to Figure 9-1 and the discussion regarding where the number 120 lies on each normal distribution in "Exploring the Basics of the Normal Distribution," you can now calculate *z*-values to get a much clearer picture. In Figure 9-1a, the number 120 is located one standard deviation above the mean, so its *z*-value is 1. In Figure 9-1b, 120 is equal to the mean, so its *z*-value is 0. Figure 9-1c shows that 120 is 3 standard deviations above the mean, so its *z*-value is 3.

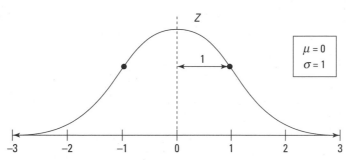

$\mu = 0$
$\sigma = 1$

High standard scores (*z*-values) aren't always the best. For example, if you're measuring the amount of time needed to run around the block, a standard score of +2 is a bad thing because your time was two standard deviations above (more than) the overall average time. In this case, a standard score of –2 would be much better, indicating your time was two standard deviations below (less than) the overall average time.

Standardizing from X to Z

Probabilities for any continuous distribution are found by finding the area under a curve (if you're into calculus, you know that means integration; if you're not into calculus, don't worry about it). Although the bell-shaped curve for the normal distribution looks easy to work with, calculating areas under its curve turns out to be a nightmare requiring high-level math procedures (believe me, I won't be going there in this book!). Plus, every normal distribution is different, causing you to repeat this process over and over each time you have to find a new probability.

To help get over this obstacle, statisticians worked out all the math gymnastics for one particular normal distribution, made a table of its probabilities, and told the rest of us to knock ourselves out. Can you guess which normal distribution they chose to crank out the table for?

Yes, all the basic results you need to find probabilities for any normal distribution (*X*) can be boiled down into one table based on the standard normal (*Z*-) distribution. This table is called the *Z*-table and is found in the appendix. Now all you need is one formula that transforms values from your normal distribution (*X*) to the *Z*-distribution; from there you can use the *Z*-table to find any probability you need.

Changing an *x*-value to a *z*-value is called *standardizing*. The so-called "*z*-formula" for standardizing an *x*-value to a *z*-value is:

$$z = \frac{x - \mu}{\sigma}$$

You take your *x*-value, subtract the mean of *X,* and divide by the standard deviation of *X.* This gives you the corresponding standard score (*z*-value or *z*-score).

Standardizing is just like changing units (for example, from Fahrenheit to Celsius). It doesn't affect probabilities for *X;* that's why you can use the *Z*-table to find them!

You can standardize an *x*-value from any distribution (not just the normal) using the *z*-formula. Similarly, not all standard scores come from a normal distribution.

Because you subtract the mean from your *x*-values and divide everything by the standard deviation when you standardize, you are literally taking the mean and standard deviation of *X* out of the equation. This is what allows you to compare everything on the scale from –3 to +3 (the *Z*-distribution) where negative values indicate being below the mean, positive values indicate being above the mean, and a value of 0 indicates you're right on the mean.

Standardizing also allows you to compare numbers from different distributions. For example, suppose Bob scores 80 on both his math exam (which has a mean of 70 and standard deviation of 10) and his English exam (which has a mean of 85 and standard deviation of 5). On which exam did Bob do better, in terms of his relative standing in the class?

Bob's math exam score of 80 standardizes to a *z*-value of $\frac{80-70}{10} = \frac{10}{10} = 1$. That tells us his math score is one standard deviation above the class average. His English exam score of 80 standardizes to a *z*-value of $\frac{80-85}{5} = \frac{-5}{5} = -1$, putting him one standard deviation below the class average. Even though Bob scored 80 on both exams, he actually did better on the math exam than the English exam, relatively speaking.

To interpret a standard score, you don't need to know the original score, the mean, or the standard deviation. The standard score gives you the relative standing of a value, which in most cases is what matters most. In fact, on most national achievement tests, they won't even tell you what the mean and standard deviation were when they report your results; they just tell you where you stand on the distribution by giving you your *z*-score.

Finding probabilities for Z with the Z-table

A full set of less-than probabilities for a wide range of *z*-values is in the *Z*-table (Table A-1 in the appendix). To use the *Z*-table to find probabilities for the standard normal (*Z*-) distribution, do the following:

1. **Go to the row that represents the first digit of your *z*-value and the first digit after the decimal point.**

2. **Go to the column that represents the second digit after the decimal point of your *z*-value.**

3. **Intersect the row and column.**

   This result represents $p(Z < z)$, the probability that the random variable Z is less than the number z (also known as the percentage of *z*-values that are less than yours).

For example, suppose you want to find $p(Z < 2.13)$. Using the Z-table, find the row for 2.1 and the column for 0.03. Intersect that row and column to find the probability: 0.9834. You find that $p(Z < 2.13) = 0.9834$.

Suppose you want to look for $p(Z < -2.13)$. You find the row for −2.1 and the column for 0.03. Intersect the row and column and you find 0.0166; that means $p(Z < -2.13)$ equals 0.0166. (This happens to be one minus the probability that Z is less than 2.13 because $p(Z < +2.13)$ equals 0.9834. That's true because the normal distribution is symmetric; more on that in the following section.)

Finding Probabilities for a Normal Distribution

Here are the steps for finding a probability when X has any normal distribution:

1. **Draw a picture of the distribution.**

2. **Translate the problem into one of the following: $p(X < a)$, $p(X > b)$, or $p(a < X < b)$. Shade in the area on your picture.**

3. **Standardize *a* (and/or *b*) to a *z*-score using the *z*-formula:**

   $$z = \frac{x - \mu}{\sigma}$$

4. **Look up the *z*-score on the Z-table (Table A-1 in the appendix) and find its corresponding probability.**

   (See the section "Standardizing from X to Z" for more on the Z-table).

5a. **If you need a "less-than" probability — that is, $p(X < a)$ — you're done.**

5b. **If you want a "greater-than" probability — that is, $p(X > b)$ — take one minus the result from Step 4.**

5c. **If you need a "between-two-values" probability — that is, $p(a < X < b)$ — do Steps 1–4 for *b* (the larger of the two values) and again for *a* (the smaller of the two values), and subtract the results.**

HEADS UP

The probability that X is equal to any single value is 0 for any continuous random variable (like the normal). That's because continuous random variables consider probability as being area under the curve, and there's no area under a curve at one single point. This isn't true of discrete random variables.

Suppose, for example, that you enter a fishing contest. The contest takes place in a pond where the fish lengths have a normal distribution with mean $\mu = 16$ inches and standard deviation $\sigma = 4$ inches.

> ✔ Problem 1: What's the chance of catching a small fish — say, less than 8 inches?
>
> ✔ Problem 2: Suppose a prize is offered for any fish over 24 inches. What's the chance of winning a prize?
>
> ✔ Problem 3: What's the chance of catching a fish between 16 and 24 inches?

To solve these problems using the steps that I just listed, first draw a picture of the normal distribution at hand. Figure 9-3 shows a picture of X's distribution for fish lengths. You can see where the numbers of interest (8, 16, and 24) fall.

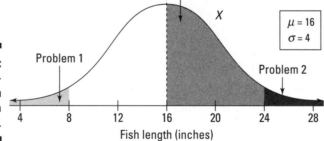

Figure 9-3: The distribution of fish lengths in a pond.

Next, translate each problem into probability notation. Problem 1 is really asking you to find $p(X < 8)$. For Problem 2, you want $p(X > 24)$. And Problem 3 is looking for $p(16 < X < 24)$.

Step 3 says change the x-values to z-values using the z-formula:

$$z = \frac{x - \mu}{\sigma}$$

For Problem 1 of the fish example, you have the following:

$$p(X < 8) = p\left(Z < \frac{8 - 16}{4}\right) = p(Z < -2)$$

Similarly for Problem 2, $p(X > 24)$ becomes

$$p(X > 24) = p\left(Z > \frac{24-16}{4}\right) = p(Z > 2)$$

And Problem 3 translates from $p(16 < X < 24)$ to

$$p(16 < X < 24) = p\left(\frac{16-16}{4} < Z < \frac{24-16}{4}\right) = p(0 < Z < 2)$$

Figure 9-4 shows a comparison of the X-distribution and Z-distribution for the values x = 8, 16, and 24, which standardize to z = –2, 0, and +2, respectively.

Now that you have changed x-values to z-values, you move to Step 4 and find (or calculate) probabilities for those z-values using the Z-table (in the appendix). In Problem 1 of the fish example, you want $p(Z < –2)$; go to the Z-table and look at the row for –2.0 and the column for 0.00, intersect them, and you find 0.0228 — according to Step 5a, you're done. The chance of a fish being less than 8 inches is equal to 0.0228.

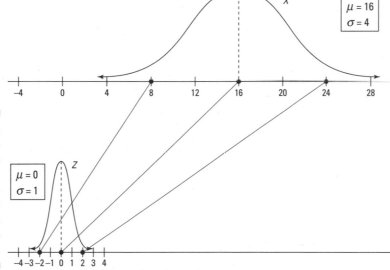

Figure 9-4:
Standardizing numbers from a normal distribution (X) to numbers on the Z-distribution.

For Problem 2, find $p(Z > 2.00)$. Because it's a "greater-than" problem, this calls for Step 5b. To be able to use the Z-table, you need to rewrite this in terms of a "less-than" statement. Because the entire probability for the Z-distribution equals 1, we know $p(Z > 2.00) = 1 - p(Z < 2.00) = 1 - 0.9772 = 0.0228$ (using the Z-table). So, the chance that a fish is greater than 24 inches is also 0.0228. (Note: The answers to Problems 1 and 2 are the same because the Z-distribution is symmetric; refer to Figure 9-3.)

In Problem 3, you find $p(0 < Z < 2.00)$; this requires Step 5c. First find $p(Z < 2.00)$, which is 0.9772 from the Z-table. Then find $p(Z < 0)$, which is 0.5000 from the Z-table. Subtract them to get $0.9772 - 0.5000 = 0.4772$. The chance of a fish being between 16 and 24 inches is 0.4772.

The Z-table does not list every possible value of Z; it just carries them out to two digits after the decimal point. Use the one closest to the one you need. And just like in an airplane where the closest exit may be behind you, the closest z-value may be the one that is lower than the one you need.

Finding X When You Know the Percent

Another popular normal distribution problem involves finding percentiles for X (see Chapter 5 for a detailed rundown on percentiles). That is, you are given the percentage or probability of being at or below a certain x-value, and you have to find the x-value that corresponds to it. For example, if you know that the people whose golf scores were in the lowest 10% got to go to the tournament, you may wonder what the cutoff score was; that score would represent the 10th percentile.

A percentile isn't a percent. A percent is a number between 0 and 100; a percentile is a value of X (a height, an IQ, a test score, and so on).

Figuring out a percentile for a normal distribution

Certain percentiles are so popular that they have their own names and their own notation. The three "named" percentiles are Q_1 — the first quartile, or the 25th percentile; Q_2 — the 2nd quartile (also known as the *median* or the 50th percentile); and Q_3 — the 3rd quartile or the 75th percentile. (See Chapter 5 for more information on quartiles.)

Here are the steps for finding any percentile for a normal distribution X:

1a. **If you're given the probability (percent) less than x and you need to find x, you translate this as: Find a where $p(X < a) = p$ (and p is the given probability). That is, find the pth percentile for X. Go to Step 2.**

1b. **If you're given the probability (percent) greater than x and you need to find x, you translate this as: Find b where $p(X > b) = p$ (and p is given). Rewrite this as a percentile (less-than) problem: Find b where $p(X < b) = 1 - p$. This means find the $(1 - p)$th percentile for X.**

2. **Find the corresponding percentile for Z by looking in the body of the Z-table (in the appendix) and finding the probability that is closest to p (from Step 1a) or 1 – p (from Step 1b). Find the row and column this probability is in (using the table backwards). This is the desired z-value.**

3. **Change the z-value back into an x-value (original units) by using x = μ + zσ. You've (finally!) found the desired percentile for X.**

   The formula in this step is just a rewriting of the z-formula, $z = \dfrac{x - \mu}{\sigma}$, so it's solved for x.

Doing a low percentile problem

Look at the fish example used previously in "Finding Probabilities for a Normal Distribution," where the lengths (X) of fish in a pond have a normal distribution with mean 16 inches and standard deviation 4 inches. Suppose you want to know what length marks the bottom 10 percent of all the fish lengths in the pond. What percentile are you looking for?

Being at the bottom 10 percent means you have a "less-than" probability that's equal to 10 percent, and you are at the 10th percentile.

Now go to Step 1a in the preceding section and translate the problem. In this case, because you're dealing with a "less-than" situation, you want to find x such that p(X < x) = 0.10. This represents the 10th percentile for X. Figure 9-5 shows a picture of this situation.

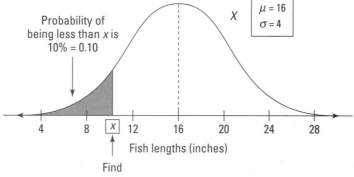

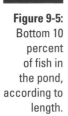

Figure 9-5: Bottom 10 percent of fish in the pond, according to length.

Now go to Step 2, which says to find the 10th percentile for Z. Looking in the body of the Z-table (in the appendix), the probability closest to 0.10 is 0.1003, which falls in the row for z = –1.2 and the column for 0.08. That means the 10th percentile for Z is –1.28; so a fish whose length is 1.28 standard deviations below the mean marks the bottom 10 percent of all fish lengths in the pond.

But exactly how long is that fish, in inches? In Step 3, you change the z-value back to an x-value (fish length in inches) using the z-formula solved for x; you get $x = 16 + -1.28 * 4 = 10.88$ inches. So 10.88 inches marks the lowest 10 percent of fish lengths. Ten percent of the fish are shorter than that.

Working with a higher percentile

Now suppose you want to find the length that marks the *top* 25 percent of all the fish in the pond. This problem calls for Step 1b (in "Finding a percentile for a normal distribution") because being in the top part of the distribution means you're dealing with a greater-than probability. The number you are looking for is somewhere in the right tail (upper area) of the X-distribution, with p = 25 percent of the probability to its right and $1 - p$ = 75 percent to its left. Thinking in terms of the Z-table and how it only uses less-than probabilities, you need to find the 75th percentile for Z, then change it to an x-value.

Step 2: The 75th percentile of Z is the z-value where $p(Z < z) = 0.75$. Using the Z-table (in the appendix), you find the probability closest to 0.7500 is 0.7486, and its corresponding z-value is in the row for 0.6 and column for 0.07. Put these together and you get a z-value of 0.67. This is the 75th percentile for Z. In Step 3, change the z-value back to an x-value (length in inches) using the z-formula solved for x to get $x = 16 + 0.67 * 4 = 18.68$ inches. So, 75% of the fish are shorter than 18.68 inches. And to answer the original question, the top 25% of the fish in the pond are longer than 18.68 inches.

Translating tricky wording in percentile problems

Some percentile problems are especially challenging to translate. For example, suppose the amount of time for a racehorse to run around a track in a qualifying round has a normal distribution with mean 120 seconds and standard deviation 5 seconds. The best 10 percent of the times qualify; the rest don't. What's the cutoff time for qualifying?

Because "best times" mean "lowest times" in this case, the percentage of times that lie *below* the cutoff must be 10, and the percentage *above* the cutoff must be 90. (It's an easy mistake to think it's the other way around.) The percentile of interest is therefore the 10th, which is down on the left tail of the distribution. You now work this problem the same way I worked Problem 1 regarding fish lengths (see the section, "Finding Probabilities for a Normal Distribution"). The standard score for the 10th percentile is $z = -1.28$ looking at the Z-table (in the appendix). Converting back to original units, you get $x = \mu + z\sigma = 120 + (-1.28)(5) = 113.6$ seconds. So the cutoff time needed for a racehorse to qualify (that is, to be among the fastest 10%) is 113.6 seconds. (Notice this number is less than the average time of 120 seconds, which makes sense; a negative z-value is what makes this happen.)

The 50th percentile for the normal distribution is the mean (because of symmetry) and its *z*-score is zero. Smaller percentiles, like the 10th, lie below the mean and have negative *z*-scores. Larger percentiles, like the 75th, lie above the mean and have positive *z*-scores.

Here's another style of wording that has a bit of a twist: Suppose times to complete a statistics exam have a normal distribution with a mean of 40 minutes and standard deviation of 6 minutes. Deshawn's time comes in at the 90th percentile. What percentage of the students are still working on their exams when Deshawn leaves? Because Deshawn is at the 90th percentile, 90 percent of the students have exam times lower than hers. That means 90% of the students left before Deshawn, so $100 - 90 = 10$ percent of the students are still working when Deshawn leaves.

To be able to decipher the language used to imply a percentile problem, look for clues like *the bottom 10%* (also known as the 10th percentile) and *the top 10%* (also known as the 90th percentile). For *the best 10%,* you must determine whether low or high numbers qualify as "best."

Normal Approximation to the Binomial

Suppose you flip a fair coin 100 times and you let *X* equal the number of heads. What's the probability that *X* is greater than 60? In Chapter 8, you solve problems like this (involving fewer flips) using the binomial distribution. For binomial problems where *n* (the number of trials) is small, you can either use the direct formula (found in Chapter 8), the binomial table (found in the appendix), or you can use technology if available (such as a graphing calculator or Microsoft Excel).

However, if *n* is large the calculations get unwieldy and the binomial table runs out of numbers. If there's no technology available (like when taking an exam), what can you do to find a binomial probability? Turns out, if *n* is large enough, you can use the normal distribution to find a very close approximate answer with a lot less work.

But what do I mean by *n* being "large enough"? To determine whether *n* is large enough to use what statisticians call the *normal approximation to the binomial,* both of the following conditions must hold:

✔ $n * p \geq 10$ (at least 10), where *p* is the probability of success

✔ $n * (1 - p) \geq 10$ (at least 10), where $1 - p$ is the probability of failure

To find the normal approximation to the binomial distribution when n is large, use the following steps:

1. **Verify whether n is large enough to use the normal approximation by checking the two appropriate conditions.**

   For the coin-flipping question, the conditions are met because $n * p = 100 * 0.50 = 50$, and $n * (1 - p) = 100 * (1 - 0.50) = 50$, both of which are at least 10. So go ahead with the normal approximation.

2. **Translate the problem into a probability statement about X.**

   For the coin-flipping example, you need to find $p(X > 60)$.

3. **Standardize the x-value to a z-value, using the z-formula:**

   $$z = \frac{x - \mu}{\sigma}$$

   For the mean of the normal distribution, use $\mu = np$ (the mean of the binomial), and for the standard deviation σ, use $\sqrt{np(1-p)}$ (the standard deviation of the binomial; see Chapter 8).

   For the coin-flipping example, use $\mu = np = (100)(0.50) = 50$ and $\sigma = \sqrt{np(1-p)} = \sqrt{100(0.50)(1-0.50)} = 5$. Then put these values into the z-formula to get $z = \frac{x - \mu}{\sigma} = \frac{60 - 50}{5} = 2$. To solve the problem, you need to find $p(Z > 2)$.

On an exam, you won't see μ and σ in the problem when you have a binomial distribution. However, you know the formulas that allow you to calculate both of them using n and p (both of which will be given in the problem). Just remember you have to do that extra step to calculate the μ and σ needed for the z-formula.

4. **Proceed as you usually would for any normal distribution. That is, do Steps 4 and 5 described in the earlier section "Finding Probabilities for a Normal Distribution."**

   Continuing the example, $p(Z > 2.00) = 1 - 0.9772 = 0.0228$ from the Z-table (appendix). So the chance of getting more than 60 heads in 100 flips of a coin is only about 2.28 percent. (I wouldn't bet on it.)

When using the normal approximation to find a binomial probability, your answer is an *approximation* (not exact) — be sure to state that. Also show that you checked both necessary conditions for using the normal approximation.

Chapter 10

The *t*-Distribution

The *t*-distribution is one of the mainstays of data analysis. You may have heard of the "*t*-test" for example, which is often used to compare two groups in medical studies and scientific experiments.

This short chapter covers the basic characteristics and uses of the *t*-distribution. You find out how it compares to the normal distribution (more on that in Chapter 9) and how to use the *t*-table to find probabilities and percentiles.

Basics of the t-Distribution

In this section, you get an overview of the *t*-distribution, its main characteristics, when it's used, and how it's related to the *Z*-distribution (see Chapter 9).

Comparing the t- and Z-distributions

The normal distribution is that well-known bell-shaped distribution whose mean is μ and whose standard deviation is σ (see Chapter 9 for more on the normal distribution). The most common normal distribution is the standard normal (also called the *Z*-distribution), whose mean is 0 and standard deviation is 1.

The *t*-distribution can be thought of as a cousin of the standard normal distribution — it looks similar in that it's centered at zero and has a basic bell-shape, but it's shorter and flatter than the *Z*-distribution. Its standard deviation is proportionally larger compared to the *Z*, which is why you see the fatter tails on each side.

Figure 10-1 compares the *t-* and standard normal (*Z-*) distributions in their most general forms.

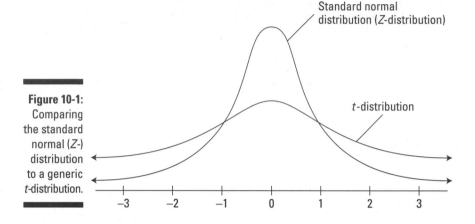

Figure 10-1:
Comparing
the standard
normal (*Z-*)
distribution
to a generic
*t-*distribution.

The *t-*distribution is typically used to study the mean of a population, rather than to study the individuals within a population. In particular, it is used in many cases when you use data to estimate the population mean — for example, to estimate the average price of all the new homes in California. Or when you use data to test someone's claim about the population mean — for example, is it true that the mean price of all the new homes in California is $500,000?

These procedures are called *confidence intervals* and *hypothesis tests* and are discussed in Chapters 13 and 14, respectively.

The connection between the normal distribution and the *t-*distribution is that the *t-*distribution is often used for analyzing the mean of a population if the population has a normal distribution (or fairly close to it). Its role is especially important if your data set is small or if you don't know the standard deviation of the population (which is often the case).

When statisticians use the term *t-distribution,* they aren't talking about just one individual distribution. There is an entire family of specific *t-*distributions, depending on what sample size is being used to study the population mean. Each *t-*distribution is distinguished by what statisticians call its *degrees of freedom.* In situations where you have one population and your sample size is *n,* the degrees of freedom for the corresponding *t-*distribution is *n* – 1. For example, a sample of size 10 uses a *t-*distribution with 10 – 1, or 9, degrees of freedom, denoted t_9 (pronounced *tee sub-nine*). Situations involving two populations use different degrees of freedom and are discussed in Chapter 15.

Discovering the effect of variability on t-distributions

t-distributions based on smaller sample sizes have larger standard deviations than those based on larger sample sizes. Their shapes are flatter; their values are more spread out. That's because results based on smaller data sets are more variable than results based on large data sets.

The larger the sample size is, the larger the degrees of freedom will be, and the more the *t*-distributions look like the standard normal distribution (*Z*-distribution). A rough cutoff point where the *t*- and *Z*-distributions become similar (at least similar enough for jazz or government work) is around $n = 30$.

Figure 10-2 shows what different *t*-distributions look like for different sample sizes and how they all compare to the standard normal (*Z*-) distribution.

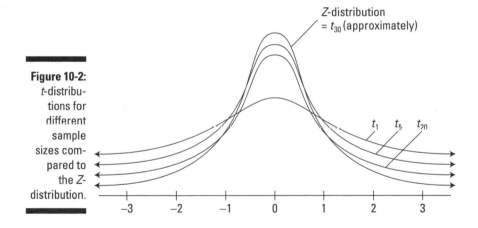

Figure 10-2: *t*-distributions for different sample sizes compared to the *Z*-distribution.

Z-distribution $= t_{30}$ (approximately)

t_1 t_5 t_{20}

Using the t-Table

Each normal distribution has its own mean and standard deviation that classify it, so finding probabilities for each normal distribution on its own is not the way to go. Thankfully, you can standardize the values of any normal distribution to become values on a standard normal (*Z*-) distribution (whose mean is 0 and standard deviation is 1) and use a *Z*-table (in the appendix) to find probabilities. (Chapter 9 has info on normal distributions.)

In contrast, a *t*-distribution is not classified by its mean and standard deviation, but by the sample size of the data set being used (*n*). Unfortunately, there is no single "standard t-distribution" that you can use to transform the numbers and find probabilities on a table. Because it wouldn't be humanly possible to create a table of probabilities and corresponding t-values for every possible t-distribution, statisticians created one table showing certain values of t-distributions for a selection of degrees of freedom and a selection of probabilities. This table is called the *t-table* (it appears in the appendix). In this section, you find out how to find probabilities, percentiles, and critical values (for confidence intervals) using the t-table.

Finding probabilities with the t-table

Each row of the t-table (in the appendix) represents a different t-distribution, classified by its degrees of freedom (*df*). The columns represent various common greater-than probabilities, such as 0.40, 0.25, 0.10, and 0.05. The numbers across a row indicate the values on the t-distribution (the t-values) corresponding to the greater-than probabilities shown at the top of the columns. Rows are arranged by degrees of freedom.

 Another term for greater-than probability is *right-tail probability,* which indicates that such probabilities represent areas on the right-most end (tail) of the t-distribution.

For example, the second row of the t-table is for the t_2 distribution (2 degrees of freedom, pronounced *tee sub-two*). You see that the second number, 0.816, is the value on the t_2 distribution whose area to its right (its right-tail probability) is 0.25 (see the heading for column 2). In other words, the probability that t_2 is greater than 0.816 equals 0.25. In probability notation, that means $p(t_2 > 0.816) = 0.25$.

The next number in row two of the t-table is 1.886, which lies in the 0.10 column. This means the probablity of being greater than 1.886 on the t_2 distribution is 0.10. Because 1.886 falls to the right of 0.816, its right-tail probability is lower.

Figuring percentiles for the t-distribution

You can also use the t-table (in the appendix) to find percentiles for a t-distribution. A *percentile* is a number on a distribution whose less-than probability is the given percentage; for example, the 95th percentile of the t-distribution with $n - 1$ degrees of freedom is that value of t_{n-1} whose left-tail (less-than) probability is 0.95 (and whose right-tail probability is 0.05). (See Chapter 5 for particulars on percentiles.)

Suppose you have a sample of size 10 and you want to find the 95th percentile of its corresponding t-distribution. You have $n - 1 = 9$ degrees of freedom, so you look at the row for *df* = 9. The 95th percentile is the number where 95% of the values lie below it and 5% lie above it, so you want the right-tail area to be 0.05. Move across the row, find the column for 0.05, and you get t_9 = 1.833. This is the 95th percentile of the *t*-distribution with 9 degrees of freedom.

Now, if you increase the sample size to *n* = 20, the value of the 95th percentile decreases; look at the row for 20 – 1 = 19 degrees of freedom, and in the column for 0.05 (a right-tail probability of 0.05) you find t_{19} = 1.729. Notice that the 95th percentile for the t_{19} distribution is less than the 95th percentile for the t_9 distribution (1.833). This is because larger degrees of freedom indicate a smaller standard deviation and the *t*-values are more concentrated about the mean, so you reach the 95th percentile with a smaller value of *t*. (See the section "Discovering the effect of variability on t-distributions," earlier in this chapter.)

Picking out t*-values for confidence intervals

Confidence intervals estimate population parameters, such as the population mean, by using a statistic (for example, the sample mean) plus or minus a margin of error. (See Chapter 13 for all the information you need on confidence intervals and more.) To compute the margin of error for a confidence interval, you need a *critical value* (the number of standard errors you add and subtract to get the margin of error you want; see Chapter 13). When the sample size is large (at least 30), you use critical values on the *Z*-distribution (shown in Chapter 13) to build the margin of error. When the sample size is small (less than 30) and/or the population standard deviation is unknown, you use the *t*-distribution to find critical values.

To help you find critical values for the *t*-distribution, you can use the last row of the *t*-table, which lists common confidence levels, such as 80%, 90%, and 95%. To find a critical value, look up your confidence level in the bottom row of the table; this tells you which column of the *t*-table you need. Intersect this column with the row for your *df* (see Chapter 13 for degrees of freedom formulas). The number you see is the critical value (or the *t**-value) for your confidence interval. For example, if you want a *t**-value for a 90% confidence interval when you have 9 degrees of freedom, go to the bottom of the table, find the column for 90%, and intersect it with the row for *df* = 9. This gives you a *t**-value of 1.833 (rounded).

Across the top row of the *t*-table, you see right-tail probabilities for the *t*-distribution. But confidence intervals involve both left- and right-tail probabilities (because you add and subtract the margin of error). So half of the probability left from the confidence interval goes into each tail. You need to take that into account. For example, a *t**-value for a 90% confidence interval has 5% for its greater-than probability and 5% for its less-than probability (taking 100% minus 90% and dividing by 2). Using the top row of the *t*-table, you would have to look for 0.05 (rather than 10%, as you might be inclined to do.) But using the bottom row of the table, you just look for 90%. (The result you get using either method ends up being in the same column.)

When looking for t^*-values for confidence intervals, use the bottom row of the t-table as your guide, rather than the headings at the top of the table.

Studying Behavior Using the t-Table

You can use computer software to calculate any probabilities, percentiles, or critical values you need for any t-distribution (or any other distribution) if it's available to you. (On exams it may not be available.) However, one of the nice things about using a table to find probabilities (rather than using computer software) is that the table can tell you information about the behavior of the distribution itself — that is, it can give you the big picture. Here are some nuggets of big-picture information about the t-distribution you can glean by scanning the t-table (in the appendix).

In Figure 10-2, as the degrees of freedom increase, the values on each t-distribution become more concentrated around the mean, eventually resembling the Z-distribution. The t-table confirms this pattern as well. Because of the way the t-table is set up, if you choose any column and move down through the numbers in the column, you're increasing the degrees of freedom (and sample size) and keeping the right-tail probability the same. As you do this, you see the t-values getting smaller and smaller, indicating the t-values are becoming closer to (hence more concentrated around) the mean.

I labeled the second-to-last row of the t-table with a z in the df column. This indicates the "limit" of the t-values as the sample size (n) goes to infinity. The t-values in this row are approximately the same as the z-values on the Z-table (in the appendix) that correspond to the same greater-than probabilities. This confirms what you already know: As the sample size increases, the t- and the Z-distributions look more and more alike. For example, the t-value in row 30 of the t-table corresponding to a right-tail probability of 0.05 (column 0.05) is 1.697. This lies close to $z = 1.645$, the value corresponding to a right-tail area of 0.05 on the Z-distribution. (See row Z of the t-table.)

It doesn't take a super-large sample size for the values on the t-distribution to get close to the values on a Z-distribution. For example, when $n = 31$ and $df = 30$, the values in the t-table are already quite close to the corresponding values on the Z-table.

Chapter 11

Sampling Distributions and the Central Limit Theorem

....

In This Chapter

▶ Understanding the concept of a sampling distribution

▶ Putting the Central Limit Theorem to work

▶ Determining the factors that affect precision

....

When you take a sample of data, it's important to realize the results will vary from sample to sample. Statistical results based on samples should include a measure of how much those results are expected to vary. When the media reports statistics like the average price of a gallon of gas in the U.S. or the percentage of homes on the market that were sold over the last month, you know they didn't sample every possible gas station or every possible home sold. The question is, how much would their results change if another sample was selected?

This chapter addresses this question by studying the behavior of means for all possible samples, and the behavior of proportions from all possible samples. By studying the behavior of all possible samples, you can gauge where your sample results fall and understand what it means when your sample results fall outside of certain expectations.

Defining a Sampling Distribution

A *random variable* is a characteristic of interest that takes on certain values in a random manner. For example, the number of red lights you hit on the way to work or school is a random variable; the number of children a randomly selected family has is a random variable. You use capital letters such as X or Y to denote random variables and you use small case letters x or y

to denote actual outcomes of random variables. A *distribution* is a listing, graph, or function of all possible outcomes of a random variable (such as X) and how often each actual outcome (x), or set of outcomes, occurs. (See Chapter 8 for more details on random variables and distributions.)

For example, suppose a million of your closest friends each rolls a single die and records each actual outcome (x). A table or graph of all these possible outcomes (one through six) and how often they occurred represents the distribution of the random variable X. A graph of the distribution of X in this case is shown in Figure 11-1a. It shows the numbers 1–6 appearing with equal frequency (each one occurring ⅙ of the time), which is what you expect over many rolls if the die is fair.

Now suppose each of your friends rolls this single die 50 times ($n = 50$) and records the average, $\bar{x}$. The graph of all their averages of all their samples represents the distribution of the random variable $\overline{X}$. Because this distribution is based on sample averages rather than individual outcomes, this distribution has a special name. It's called the *sampling distribution* of the sample mean, $\overline{X}$. Figure 11-1b shows the sampling distribution of $\overline{X}$, the average of 50 rolls of a die.

Figure 11-1b (average of 50 rolls) shows the same range (1 through 6) of outcomes as Figure 11-1a (individual rolls), but Figure 11-1b has more possible outcomes. You could get an average of 3.3 or 2.8 or 3.9 for 50 rolls, for example, whereas someone rolling a single die can only get whole numbers from 1 to 6. Also, the shape of the graphs are different; Figure 11-1a shows a flat shape, where each outcome is equally likely, and Figure 11-1b has a mound shape; that is, outcomes near the center (3.5) occur with high frequency and outcomes near the edges (1 and 6) occur with extremely low frequency. A detailed look at the differences and similarities in shape, center, and spread for individuals versus averages, and the reasons behind them, is the topic of the following sections. (See Chapter 8 if you need background info on shape, center, and spread of random variables before diving in.)

The Mean of a Sampling Distribution

Using the die-rolling example from the preceding section, X is a random variable denoting the outcome you can get from a single die (assuming the die is fair). The mean of X (over all possible outcomes) is denoted by μ_x (pronounced *mu sub-x*); in this case its value is 3.5 (as shown in Figure 11-1a). If you roll a die 50 times and take the average, the random variable $\overline{X}$ represents any outcome you could get. The mean of $\overline{X}$, denoted $\mu_{\bar{x}}$ (pronounced *mu sub-x-bar*) equals 3.5 as well. (You can see this result in Figure 11-1b.)

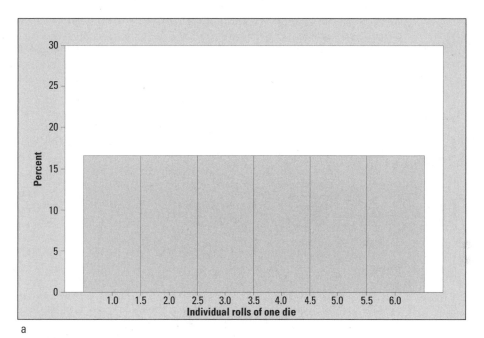

a

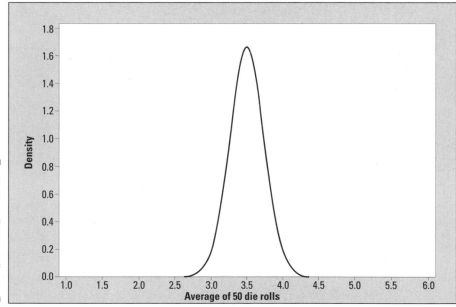

Figure 11-1:
Distributions
of a) individ-
ual rolls of
one die; and
b) average
of 50 rolls of
one die.

b

This result is no coincidence! In general, the mean of the population of all possible sample means is the same as the mean of the original population. (Notationally speaking, you write $\mu_{\bar{x}} = \mu_x$.) It's a mouthful, but it makes sense that the average of the averages from all possible samples is the same as the average of the population that the samples came from. In the die rolling example, the average of the population of all 50-roll averages equals the average of the population of all single rolls (3.5).

Using subscripts on μ, you can distinguish which mean you're talking about — the mean of X (all individuals in a population) or the mean of $\overline{X}$ (all sample means from the population).

Measuring Standard Error

The values in any population deviate from their mean; for instance, people's heights differ from the overall average height. Variability in a population of individuals (X) is measured in *standard deviations* (see Chapter 5 for details on standard deviation). Sample means vary because you're not sampling the whole population, only a subset; and as samples vary, so will their means. Variability in the sample mean ($\overline{X}$) is measured in terms of *standard errors*.

Error here doesn't mean there's been a mistake — it means there is a gap between the population and sample results.

The standard error of the sample mean is denoted by $\sigma_{\bar{x}}$ *(sigma sub-x-bar)*. Its formula is $\dfrac{\sigma_x}{\sqrt{n}}$, where σ_x is population standard deviation *(sigma sub-x)* and n is size of each sample. In the next sections you see the effect each of these two components has on the standard error.

Sample size and standard error

The first component of standard error is the sample size, n. Because n is in the denominator of the standard error formula, the standard error decreases as n increases. It makes sense that having more data gives less variation (and more precision) in your results.

Suppose X is the time it takes for a clerical worker to type and send one letter of recommendation, and say X has a normal distribution with mean 10.5 minutes and standard deviation 3 minutes. The bottom curve in Figure 11-2 shows the picture of the distribution of X, the individual times for all clerical workers in the population. According to the Empirical Rule (see Chapter 9), most of the values are within 3 standard deviations of the mean (10.5) — between 1.5 and 19.5.

Now take a random sample of 10 clerical workers, measure their times, and find the average, $\bar{x}$, each time. Repeat this process over and over, and graph all the possible results for all possible samples. The middle curve in Figure 11-2 shows the picture of the sampling distribution of $\overline{X}$. Notice that it's still centered at 10.5 (which you expected) but its variability is smaller; the standard error in this case is $\frac{\sigma_x}{\sqrt{n}} = \frac{3}{\sqrt{10}} = 0.95$ minutes (quite a bit less than 3 minutes, the standard deviation of the individual times).

Looking at Figure 11-2, the average times for samples of 10 clerical workers are closer to the mean (10.5) than the individual times are. That's because average times don't change as much from sample to sample as individual times change from person to person.

Now take all possible random samples of 50 clerical workers and find their means; the sampling distribution is shown in the tallest curve in Figure 11-2. The standard error of $\overline{X}$ goes down to $\frac{\sigma_x}{\sqrt{n}} = \frac{3}{\sqrt{50}} = 0.42$ minutes. You can see the average times for 50 clerical workers are even closer to 10.5 than the ones for 10 clerical workers. By the Empirical Rule, most of the values fall between $10.5 - 3(.42) = 9.24$ and $10.5 + 3(.42) = 11.76$. Larger samples give even more precision around the mean because they change even less from sample to sample.

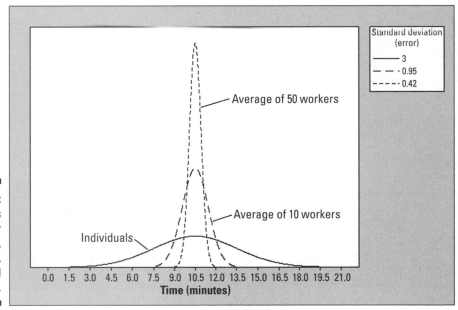

Figure 11-2:
Distributions of times for 1 worker, 10 workers, and 50 workers.

Why is having more precision around the mean important? Because sometimes you don't know the mean but want to determine what it is, or at least get as close to it as possible. How can you do that? By taking a large random sample from the population and finding its mean. You know that your sample mean will be close to the actual population mean if your sample is large, as Figure 11-2 shows (assuming your data are collected correctly; see Chapter 16 for details on collecting good data).

Population standard deviation and standard error

The second component of standard error involves the amount of diversity in the population (measured by standard deviation). In the standard error formula $\frac{\sigma_x}{\sqrt{n}}$, for $\overline{X}$, you see the population standard deviation, σ_x, is in the numerator. That means as the population standard deviation increases, the standard error of the sample means also increases. Mathematically this makes sense; how about statistically?

Suppose you have two ponds full of fish (call them pond #1 and pond #2), and you're interested in the length of the fish in each pond. Assume the fish lengths in each pond have a normal distribution (see Chapter 9). You've been told that the fish lengths in pond #1 have a mean of 20 inches and a standard deviation of 2 inches (see Figure 11-3a). Suppose the fish in pond #2 also average 20 inches but have a larger standard deviation of 5 inches (see Figure 11-3b).

Comparing Figures 11-3a and 11-3b, you see the lengths for the two populations of fish have the same shape and mean, but the distribution in Figure 11-3b (for pond #2) has more spread, or variability, than the distribution shown in Figure 11-3a (for pond #1). This spread confirms that the fish in pond #2 vary more in length than those in pond #1.

Now suppose you take a random sample of 100 fish from pond #1, find the mean length of the fish, and repeat this process over and over. Then you do the same with pond #2. Because the lengths of individual fish in pond #2 have more variability than the lengths of individual fish in pond #1, you know the average lengths of samples from pond #2 will have more variability than the average lengths of samples from pond #1 as well. (In fact, you can calculate their standard errors using the formula earlier in this section to be 0.20 and 0.50, respectively.)

Estimating the population average is harder when the population varies a lot to begin with — estimating the population average is much easier when the population values are more consistent. The bottom line is the standard error of the sample mean is larger when the population standard deviation is larger.

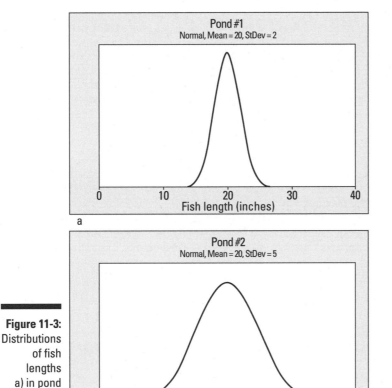

Figure 11-3:
Distributions
of fish
lengths
a) in pond
#1; b) in
pond #2.

Looking at the Shape of a Sampling Distribution

Now that you know about the mean and standard error of $\overline{X}$, the next step is to determine the shape of the sampling distribution of $\overline{X}$; that is, the shape of the distribution of all possible sample means (all possible values of $\overline{x}$) from all possible samples. You proceed differently for different conditions, which I divide into two cases: 1) the original distribution for X (the population) is normal, or has a normal distribution; and 2) the original distribution for X (the population) is *not* normal, or is unknown.

Case 1: The distribution of X is normal

If X has a normal distribution, then $\overline{X}$ does too, no matter what the sample size n is. In the example regarding the amount of time (X) for a clerical worker to complete a task (refer to the section "Sample size and standard error"), you knew X had a normal distribution (refer to the lowest curve in Figure 11-2). If you refer to the other curves in Figure 11-2, you see the average times for samples of $n = 10$ and $n = 50$ clerical workers, respectively, also have normal distributions.

When X has a normal distribution, the sample means also always have a normal distribution, no matter what size samples you take, even if you take samples of only 2 clerical workers at a time.

The difference between the curves in Figure 11-2 is not their means or their shapes, but rather their amount of variability (how close the values in the distribution are to the mean). Results based on large samples vary less and will be more concentrated around the mean than results from small samples or results from the individuals in the population.

Case 2: The distribution of X is not normal — enter the Central Limit Theorem

If X has any distribution that is *not* normal, or if its distribution is unknown, you can't automatically say the sample mean ($\overline{X}$) has a normal distribution. But incredibly, you can use a normal distribution to *approximate* the distribution of $\overline{X}$ — if the sample size is large enough. This momentous result is due to what statisticians know and love as the Central Limit Theorem.

The *Central Limit Theorem* (abbreviated *CLT*) says that if X does *not* have a normal distribution (or its distribution is unknown and hence can't be deemed to be normal), the shape of the sampling distribution of $\overline{X}$ is *approximately* normal, as long as the sample size, n, is large enough. That is, you get an *approximate* normal distribution for the means of large samples, even if the distribution of the original values (X) is *not* normal.

Most statisticians agree that if n is at least 30, this approximation will be reasonably close in most cases, although different distribution shapes for X have different values of n that are needed. The larger the sample size (n), the closer the distribution of the sample means will be to a normal distribution.

Averaging a fair die is approximately normal

Consider the die rolling example from the earlier section "Defining a Sampling Distribution." Notice in Figure 11-1a, the distribution of X (the population of outcomes based on millions of single rolls) is flat; the individual outcomes of each roll go from 1 to 6, and each outcome is equally likely.

Things change when you look at averages. When you roll a die a large number of times (say a sample of 50 times) and look at your outcomes, you'll probably find about the same number of 6s as 1s (note that 6 and 1 average out to 3.5); 5s as 2s (5 and 2 also average out to 3.5); and 4s as 3s (which also average out to 3.5 — do you see a pattern here?). So if you roll a die 50 times, you have a high probability of getting an overall average that's close to 3.5. Sometimes just by chance things won't even out as well, but that won't happen very often with 50 rolls.

Getting an average at the extremes with 50 rolls is a very rare event. To get an average of 1 on 50 rolls, you need all 50 rolls to be 1. How likely is that? (If it happens to you, buy a lottery ticket right away, it's the luckiest day of your life!) The same is true for getting an average near 6.

So the chance that your average of 50 rolls is close to the middle (3.5) is highest, and the chance of it being at or close to the extremes (1 or 6) is extremely low. As for averages between 1 and 6, the probabilities get smaller as you move farther from 3.5, and the probabilities get larger as you move closer to 3.5; in particular, statisticians show that the shape of the sampling distribution of sample means in Figure 11-1b is *approximately* normal as long as the sample size is large enough. (See Chapter 9 for particulars on the shape of the normal distribution.)

Note that if you roll the die even more times, the chance of the average being close to 3.5 increases, and the sampling distribution of the sample means looks more and more like a normal distribution.

Averaging an unfair die is still approximately normal

However, sometimes the values of X don't occur with equal probability like they do when you roll a fair die. What happens then? For example, say the die isn't fair, and the average value for many individual rolls turns out to be 2 instead of 3.5. This means the distribution of X is skewed right (more low values like 1, 2, and 3, and fewer high values like 4, 5, and 6). But if the distribution of X (millions of individual rolls of this unfair die) is skewed right, how does the distribution of $\overline{X}$ (average of 50 rolls of this unfair die) end up with an approximate normal distribution?

Say that one person, Bob, is doing 50 rolls. What will the distribution of Bob's outcomes look like? Bob is more likely to get low outcomes (like 1 and 2) and less likely to get high outcomes (like 5 and 6) — the distribution of Bob's outcomes will be skewed right as well.

In fact, because Bob rolled his die a large number of times (50), the distribution of his individual outcomes has a good chance of matching the distribution of X (the outcomes from millions of rolls). However, if Bob had only rolled his die a few times (say, 6 times), he would be unlikely to even get the higher numbers like 5 and 6, and hence his distribution wouldn't look as much like the distribution of X.

If you run through the results of each of a million people like Bob who rolled this unfair die 50 times, each of their million distributions will look very similar to each other and very similar to the distribution of X. The more rolls they make each time, the closer their distributions get to the distribution of X and to each other. And here is the key: If their distributions of outcomes have a similar shape, no matter what that similar shape is, then their averages will be similar as well. Some people will get higher averages than 2 by chance, and some will get lower averages by chance, but these types of averages get less and less likely the farther you get from 2. This means you're getting an *approximate* normal distribution centered at 2.

The big deal is, it doesn't matter if you started out with a skewed distribution, or some totally wacky distribution for X. Because each of them had a large sample size (number of rolls), the distributions of each person's sample results end up looking similar, so their averages will be similar, close together, and close to a normal distribution. In fancy lingo, the distribution of $\overline{X}$ is *approximately* normal as long as n is large enough. This is all due to the Central Limit Theorem.

In order for the CLT to work when X does *not* have a normal distribution, each person needs to roll their die enough times (that is, n must be large enough) so they have a good chance of getting all possible values of X, especially those outcomes that won't occur as often. If n is too small, some folks will not get the outcomes that have low probabilities and their means will differ from the rest by more than they should. As a result, when you put all the means together, they may not congregate around a single value. In the end, the approximate normal distribution may not show up.

Clarifying three major points about the CLT

I want to alert you to a few sources of confusion about the Central Limit Theorem before they happen to you:

✔ The CLT is needed only when the distribution of X is not a normal distribution or is unknown. It is *not* needed if X started out with a normal distribution.

✔ The formulas for the mean and standard error of $\overline{X}$ are *not* due to the CLT. These are just mathematical results that are always true. To see these formulas, check out the sections "The Mean of a Sampling Distribution" and "Measuring Standard Error," earlier in this chapter.

✔ The n stated in the CLT refers to the size of the sample you take each time, *not* the number of samples you take. Bob rolling a die 50 times is one sample of size 50, so $n = 50$. If 10 people do it, you have 10 samples, each of size 50, and n is still 50.

Finding Probabilities for the Sample Mean

After you've established through the conditions addressed in case 1 or case 2 (see the previous sections) that $\overline{X}$ has a normal or *approximately* normal distribution, you're in luck. The normal distribution is a very friendly distribution that has a table for finding probabilities and anything else you need. For example, you can find probabilities for $\overline{X}$ by converting the $\bar{x}$-value to a z-value and finding probabilities using the Z-table (provided in the appendix). (See Chapter 9 for all the details on the normal and Z-distributions.)

The general conversion formula from $\bar{x}$-values to z-values is:

$$z = \frac{\bar{x} - \mu_{\bar{x}}}{\sigma_{\bar{x}}}$$

Substituting the appropriate values of the mean and standard error of $\overline{X}$, the conversion formula becomes:

$$z = \frac{\bar{x} - \mu_x}{\sigma_x / \sqrt{n}}$$

Don't forget to divide by the square root of n in the denominator of z. Always divide by square root of n when the question refers to the *average* of the x-values.

Revisiting the clerical worker example from the previous section "Sample size and standard error," suppose X is the time it takes a randomly chosen clerical worker to type and send a standard letter of recommendation. Suppose X has a normal distribution, and assume the mean is 10.5 minutes and the standard deviation 3 minutes. You take a random sample of 50 clerical workers and measure their times. What is the chance that their average time is less than 9.5 minutes?

This question translates to finding $P(\overline{X} < 9.5)$. As X has a normal distribution to start with, you know $\overline{X}$ also has an exact (not approximate) normal distribution. Converting to z, you get:

$$z = \frac{\overline{x} - \mu_x}{\sigma_x / \sqrt{n}} = \frac{9.5 - 10.5}{3 / \sqrt{50}} = -2.36$$

So you want $P(Z < -2.36)$, which equals 0.0091 (from the Z-table in the appendix). So the chance that a random sample of 50 clerical workers average less than 9.5 minutes to complete this task is 0.91% (very small).

How do you find probabilities for $\overline{X}$ if X is *not* normal, or unknown? As a result of the CLT , the distribution of X can be non-normal or even unknown and as long as n is large enough, you can still find *approximate* probabilities for $\overline{X}$ using the standard normal (Z-)distribution and the process described earlier. That is, convert to a z-value and find approximate probabilities using the Z-table (in the appendix).

When you use the CLT to find a probability for $\overline{X}$ (that is, when the distribution of X is *not* normal or is unknown), be sure to say that your answer is an *approximation*. You also want to say the approximate answer should be close because you've got a large enough n to use the CLT. (If n is not large enough for the CLT, you can use the t-distribution in many cases — see Chapter 10.)

Beyond actual calculations, probabilities about $\overline{X}$ can help you decide whether an assumption or a claim about a population mean is on target, based on your data. In the clerical workers example, it was assumed that the average time for all workers to type up a recommendation letter was 10.5 minutes. Your sample averaged 9.5 minutes. Because the probability that they would average less than 9.5 minutes was found to be tiny (0.0091), you either got an unusually high number of fast workers in your sample just by chance, or the assumption that the average time for all workers is 10.5 minutes was simply too high. (I'm betting on the latter.) The process of checking assumptions or challenging claims about a population is called hypothesis testing; details are in Chapter 14.

The Sampling Distribution of the Sample Proportion

The Central Limit Theorem (CLT) doesn't apply only to sample means for numerical data. You can also use it with other statistics, including sample proportions for categorical data (see Chapter 6). The *population proportion, p,* is the proportion of individuals in the population who have a certain characteristic of interest (for example, the proportion of all Americans who are registered voters, or the proportion of all teenagers who own cellphones). The *sample proportion,* denoted $\hat{p}$ (pronounced *p-hat*), is the proportion of individuals in the sample who have that particular characteristic; in other words, the number of individuals in the sample who have that characteristic of interest divided by the total sample size (*n*).

For example, if you take a sample of 100 teens and find 60 of them own cellphones, the sample proportion of cellphone-owning teens is $\hat{p} = \frac{60}{100} = 0.60$. This section examines the sampling distribution of all possible sample proportions, $\hat{p}$, from samples of size *n* from a population.

The sampling distribution of $\hat{p}$ has the following properties:

✔ Its mean, denoted by $\mu_{\hat{p}}$ (pronounced *mu sub-p-hat*), equals the population proportion, *p*.

✔ Its standard error, denoted by $\sigma_{\hat{p}}$ (say *sigma sub-p-hat*), equals:

$$\sqrt{\frac{p(1-p)}{n}}$$

(Note that because *n* is in the denominator, the standard error decreases as *n* increases.)

✔ Due to the CLT, its shape is *approximately* normal, provided that the sample size is large enough. Therefore you can use the normal distribution to find approximate probabilities for $\hat{p}$.

✔ The larger the sample size (*n*), the closer the distribution of the sample proportion is to a normal distribution.

 If you are interested in the number (rather than the proportion) of individuals in your sample with the characteristic of interest, you use the binomial distribution to find probabilities for your results (see Chapter 8).

 How large is large enough for the CLT to work for sample proportions? Most statisticians agree that both *np* and *n*(1 – *p*) should be greater than or equal to 10. That is, the average number of successes (*np*) and the average number of failures *n*(1 – *p*) needs to be at least 10.

To help illustrate the sampling distribution of the sample proportion, consider a student survey that accompanies the ACT test each year asking whether the student would like some help with math skills. Assume (through past research) that 38% of all the students taking the ACT respond yes. That means p, the population proportion, equals 0.38 in this case. The distribution of responses (yes, no) for this population are shown in Figure 11-4 as a bar graph (see Chapter 6 for information on bar graphs).

Because 38% applies to all students taking the exam, I use p to denote the population proportion, rather than $\hat{p}$, which denotes sample proportions. Typically p is unknown, but I'm giving it a value here to point out how the sample proportions from samples taken from the population behave in relation to the population proportion.

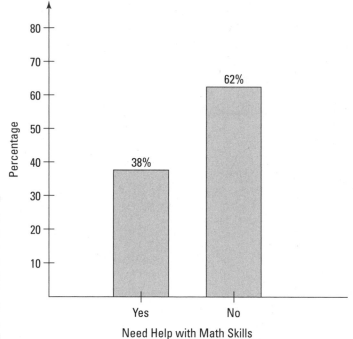

Figure 11-4:
Population percentages for responses to ACT math-help question.

Now take all possible samples of n = 1,000 students from this population and find the proportion in each sample who said they need math help. The distribution of these sample proportions is shown in Figure 11-5. It has an *approximate* normal distribution with mean p = 0.38 and standard error equal to:

$$\sqrt{\frac{p(1-p)}{n}} = \sqrt{\frac{0.38(1-0.38)}{1,000}} = 0.015$$

(or about 1.5%).

The *approximate* normal distribution works because the two conditions for the CLT are met: 1) $np = 1{,}000(0.38) = 380 \; (\geq 10)$; and 2) $n(1-p) = 1{,}000(0.62) = 620$ (also ≥ 10). And because n is so large (1,000), the approximation is excellent.

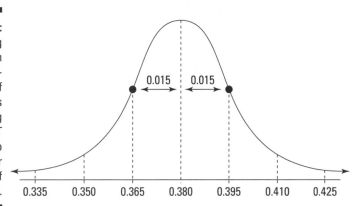

Figure 11-5:
Sampling
distribution
of propor-
tion of
students
responding
yes to ACT
math-help
question for
samples of
size 1,000.

Finding Probabilities for the Sample Proportion

You can find probabilities for $\hat{p}$, the sample proportion, by using the normal approximation as long as the conditions are met (see the previous section for those conditions). For the ACT test example, you assume that 0.38 or 38% of all the students taking the ACT test would like math help. Suppose you take a random sample of 100 students. What is the chance that more than 45 of them say they need math help? In terms of proportions, this is equivalent to the chance that more than $45 \div 100 = 0.45$ of them say they need help; that is, $P(\hat{p} > 0.45)$.

To answer this question, you first check the conditions: First, is np at least 10? Yes, because $100 * 0.38 = 38$. Next, is $n(1-p)$ at least 10? Again yes, because $100 * (1-0.38) = 62$ checks out. So you can go ahead and use the normal approximation.

You make the conversion of the $\hat{p}$-value to a z-value using the following general equation:

$$z = \frac{\hat{p} - \mu_{\hat{p}}}{\sigma_{\hat{p}}} = \frac{\hat{p} - p}{\sqrt{\dfrac{p(1-p)}{n}}}$$

When you plug in the numbers for this example, you get:

$$z = \frac{0.45 - 0.38}{\sqrt{\dfrac{0.38(1-0.38)}{100}}} = 1.44$$

And then you find $P(Z > 1.44) = 1 - 0.9251 = 0.0749$ using Table A-1 in the appendix. So if it's true that 0.38 percent of all students taking the exam want math help, the chance of taking a random sample of 100 students and finding more than 45 needing math help is *approximately* 0.0749 (by the CLT).

As noted in the previous section on sample means, you can use sample proportions to check out a claim about a population proportion. (This procedure is a hypothesis test for a population proportion; all the details are found in Chapter 15.) In the ACT example, the probability that more than 45% of the students in a sample of 100 need math help (when you assumed 38% of the population needed math help) was found to be 0.0749. Because this probability is higher than 0.05 (the typical cutoff for blowing the whistle on a claim about a population value), you can't dispute their claim that the percentage in the population needing math help is only 38%. Our sample result is just not a rare enough event. (See Chapter 15 for more on hypothesis testing for a population proportion.)

Part IV

Guesstimating and Hypothesizing with Confidence

The 5th Wave   By Rich Tennant

"What do you mean I don't fit your desired sample population at this time?"

In this part . . .

Anytime you're given a statistic by itself, you haven't really gotten the full story. The statistic alone is missing the most important part: by how much that statistic is expected to vary. All good estimates of population parameters contain not just a statistic but also a margin of error. This combination of a statistic plus or minus a margin of error is called a confidence interval.

Now suppose you're already given a claim, assumption, or target value for the population parameter, and you want to test that claim. You do it with a hypothesis test based on sample statistics. Because sample statistics will vary, you need techniques that take this into account.

This part gives you a general, intuitive look at margin of error, confidence intervals, and hypothesis tests: their function, formulas, calculations, influential factors, and interpretation. You also get quick references and examples for the most commonly used confidence intervals and hypothesis tests.

Chapter 12

Leaving Room for a Margin of Error

Good survey and experiment researchers always include some measure of how accurate their results are so that consumers of the information can put the results into perspective. This measure is called the *margin of error (MOE)* — it's a measure of how close the sample statistic (one number that summarizes the sample) is expected to be to the population parameter being studied. (A population parameter is one number that summarizes the population. Find out more about statistics and parameters in Chapter 4.) Thankfully, many journalists are also realizing the importance of the MOE in assessing information, so reports that include the margin of error are beginning to appear in the media. But what does the margin of error really mean, and does it tell the whole story?

This chapter looks at the margin of error and what it can and can't do to help you assess the accuracy of statistical information. It also examines the issue of sample size; you may be surprised at how small a sample can be used to get a good handle on the pulse of America — or the world — if the research is done correctly.

Seeing the Importance of That Plus or Minus

Margin of error is probably not a new term to you. You've probably heard of it before, most likely in the context of survey results. For example, you may have heard someone report, "This survey had a margin of error of plus or minus three percentage points." And you may have wondered what you're supposed to do with that information and how important it really is. The truth is, the survey results themselves (with no MOE) are only a measure of how the *sample* of selected individuals felt about the issue; they don't reflect how the *entire population* may have felt, had they *all* been asked. The margin of error helps you estimate how close you are to the truth about the population based on your sample data.

Results based on a sample won't be exactly the same as what you would've found for the entire population, because when you take a sample, you don't get information from everyone in the population. However, if the study is done right (see Chapters 16 and 17 for more about designing good studies), the results from the sample should be close to and representative of the actual values for the entire population, with a high level of confidence.

The MOE doesn't mean someone made a mistake; all it means is that you didn't get to sample everybody in the population, so you expect your sample results to vary from that population by a certain amount. In other words, you acknowledge that your results will change with subsequent samples and are only accurate to within a certain range — which can be calculated using the margin of error.

Consider one example of the type of survey conducted by some of the leading polling organizations, such as the Gallup Organization. Suppose its latest poll sampled 1,000 people from the United States, and the results show that 520 people (52%) think the president is doing a good job, compared to 48% who don't think so. Suppose Gallup reports that this survey had a margin of error of plus or minus 3%. Now, you know that the majority (more than 50%) of the people in this *sample* approve of the president, but can you say that the majority of *all Americans* approve of the president? In this case, you can't. Why not?

You need to include the margin of error (in this case, 3%) in your results. If 52% of *those sampled* approve of the president, you can expect that the percent of the *population of all Americans* who approve of the president will be 52%, plus or minus 3%. Therefore, between 49% and 55% of all Americans approve of the president. That's as close as you can get with your sample of 1,000. But notice that 49%, the lower end of this range, represents a minority, because it's less than 50%. So you really can't say that a majority of the American people support the president, based on this sample. You can only say you're confident that between 49% and 55% of all Americans support the president, which may or may not be a majority.

Think about the sample size for a moment. Isn't it interesting that a sample of only 1,000 Americans out of a population of well over 310,000,000 can lead you to be within plus or minus only 3% on your survey results? That's incredible! That means for large populations you only need to sample a tiny portion of the total to get close to the true value (assuming, as always, that you have good data). Statistics is indeed a powerful tool for finding out how people feel about issues, which is probably why so many people conduct surveys and why you're so often bothered to respond to them as well.

When you are working with categorical variables (those that record certain characteristics that don't involve measurements or counts; see Chapter 6), a quick-and-dirty way to get a rough idea of the margin of error for proportions, for any given sample size (n), is simply to find 1 divided by the square root of n. For the Gallup poll example, $n = 1,000$, and its square root is roughly 31.62, so the margin of error is roughly 1 divided by 31.62, or about 0.03, which is equivalent to 3%. In the remainder of this chapter, you see how to get a more accurate measure of the margin of error.

Finding the Margin of Error: A General Formula

The margin of error is the amount of "plus or minus" that is attached to your sample result when you move from discussing the sample itself to discussing the whole population that it represents. Therefore, you know that the general formula for the margin of error contains a "±" in front of it. So, how do you come up with that plus or minus amount (other than taking a rough estimate, as shown above)? This section shows you how.

Measuring sample variability

Sample results vary, but by how much? According to the Central Limit Theorem (see Chapter 11), when sample sizes are large enough, the so-called sampling distribution of the sample proportions (or the sample means) follows a bell-shaped curve (or approximate normal distribution — see Chapter 9). Some of the sample proportions (or sample means) overestimate the population value and some underestimate it, but most are close to the middle.

And what's in the middle of this sampling distribution? If you average out the results from all the possible samples you could take, the average is the actual *population proportion,* in the case of categorical data, or the actual *population average,* in the case of numerical data. Normally, you don't know all the values of the population, so you can't look at all of the possible sample results and average them out — but knowing something about all the other sample possibilities does help you to measure the amount by which you expect your own sample proportion (or average) to vary. (See Chapter 11 for more on sample means and proportions.)

Standard errors are the basic building blocks of the margin of error. The *standard error* of a statistic is basically equal to the standard deviation of the population divided by the square root of n (the sample size). This reflects the fact that the sample size greatly affects how much that sample statistic is going to vary from sample to sample. (See Chapter 11 for more about standard errors.)

The number of standard errors you have to add or subtract to get the MOE depends on how confident you want to be in your results (this is called your *confidence level*). Typically, you want to be about 95% confident, so the basic rule is to add or subtract about 2 standard errors (1.96, to be exact) to get the MOE (you get this from the Empirical Rule; see Chapter 9). This allows you to account for about 95% of all possible results that may have occurred with repeated sampling. To be 99% confident, you add and subtract 2.58 standard errors. (This assumes a normal distribution on large *n*; standard deviation known. See Chapter 11.)

You can be more precise about the number of standard errors you have to add or subtract in order to calculate the MOE for any confidence level; if the conditions are right, you can use values on the standard normal (Z-) distribution. (See Chapter 13 for details.) For any given confidence level, a corresponding value on the standard normal distribution (called a z^*-*value*) represents the number of standard errors to add and subtract to account for that confidence level. For 95% confidence, a more precise z^*-value is 1.96 (which is "about" 2), and for 99% confidence, the exact z^*-value is 2.58. Some of the more commonly used confidence levels (also known as percentage confidence), along with their corresponding z^*-values, are given in Table 12-1.

Table 12-1	z^*-Values for Selected (Percentage) Confidence Levels
Percentage Confidence	**z^*-Value**
80	1.28
90	1.645
95	1.96
98	2.33
99	2.58

To find a z^*-value like those in Table 12-1, add to the confidence level to make it a less-than probability and find its corresponding z-value on the Z-table. For example, a 95% confidence level means the "between" probability is 95%, so the "less-than" probability is 95% plus 2.5% (half of what's left), or 97.5%. Look up 0.975 in the body of the Z-table and find $z^* = 1.96$ for a 95% confidence level.

Calculating margin of error for a sample proportion

When a polling question asks people to choose from a range of answers (for example, "Do you approve or disapprove the president's performance?"), the statistic used to report the results is the proportion of people from the sample who fell into a certain group (for example, the "approve" group). This is known as the *sample proportion*. You find this number by taking the number of people in the sample that fell into the group of interest, divided by the sample size, n.

Along with the sample proportion, you need to report a margin of error. The general formula for margin of error for the sample proportion (if certain conditions are met) is $z*\sqrt{\frac{\hat{p}(1-\hat{p})}{n}}$, where $\hat{p}$ is the sample proportion, n is the sample size, and $z*$ is the appropriate $z*$-value for your desired level of confidence (from Table 12-1). Here are the steps for calculating the margin of error for a sample proportion:

1. **Find the sample size, n, and the sample proportion, $\hat{p}$.**

   The sample proportion is the number in the sample with the characteristic of interest, divided by n.

2. **Multiply the sample proportion by $(1-\hat{p})$.**

3. **Divide the result by n.**

4. **Take the square root of the calculated value.**

   You now have the standard error, $\sqrt{\frac{\hat{p}(1-\hat{p})}{n}}$.

5. **Multiply the result by the appropriate $z*$-value for the confidence level desired.**

   Refer to Table 12-1 for the appropriate $z*$-value. If the confidence level is 95%, the $z*$-value is 1.96.

Looking at the example involving whether Americans approve of the president, you can find the actual margin of error. First, assume you want a 95% level of confidence, so $z* = 1.96$. The number of Americans in the sample who said they approve of the president was found to be 520. This means that the sample proportion, $\hat{p}$, is $520 \div 1{,}000 = 0.52$. (The sample size, n, was 1,000.) The margin of error for this polling question is calculated in the following way:

$$z*\sqrt{\frac{\hat{p}(1-\hat{p})}{n}} = 1.96\sqrt{\frac{(0.52)(0.48)}{1{,}000}}$$

$$= (1.96)(0.0158) = 0.0310$$

According to this data, you conclude with 95% confidence that 52% of all Americans approve of the president, plus or minus 3.1%.

Two conditions need to be met in order to use a $z*$-value in the formula for margin of error for a sample proportion:

1. You need to be sure that $n\hat{p}$ is at least 10.

2. You need to make sure that $n(1-\hat{p})$ is at least 10.

In the preceding example of a poll on the president, $n = 1,000$, $\hat{p} = 0.52$, and $1-\hat{p}$ is $1 - 0.52 = 0.48$. Now check the conditions: $n\hat{p} = 1,000 * 0.52 = 520$, and $n(1-\hat{p}) = 1,000 * 0.48 = 480$. Both of these numbers are at least 10, so everything is okay.

Most surveys you come across are based on hundreds or even thousands of people, so meeting these two conditions is usually a piece of cake (unless the sample proportion is very large or very small, requiring a larger sample size to make the conditions work).

A sample proportion is the decimal version of the sample percentage. In other words, if you have a sample percentage of 5%, you must use 0.05 in the formula, not 5. To change a percentage into decimal form, simply divide by 100. After all your calculations are finished, you can change back to a percentage by multiplying your final answer by 100%.

Reporting results

Including the margin of error allows you to make conclusions beyond your sample to the population. After you calculate and interpret the margin of error, report it along with your survey results. To report the results from the president approval poll in the previous section, you say, "Based on my sample, 52% of all Americans approve of the president, plus or minus a margin of error of 3.1%. I am 95% confident in these results."

How does a real-life polling organization report its results? Here's an example from Gallup:

> *Based on the total random sample of 1,000 adults in (this) survey, we are 95% confident that the margin of error for our sampling procedure and its results is no more than ±3.1 percentage points.*

It sounds sort of like that long list of disclaimers that comes at the end of a car-leasing advertisement. But now you can understand the fine print!

Never accept the results of a survey or study without the margin of error for the study. The MOE is the only way to estimate how close the sample statistics are to the actual population parameters you're interested in. Sample results vary, and if a different sample had been chosen, a different sample result would have been obtained; the MOE measures that amount of difference.

The next time you hear a media story about a survey or poll that was conducted, take a closer look to see if the margin of error is given; if it's not, you should ask why. Some news outlets are getting better about reporting the margin of error for surveys, but what about other studies?

Calculating margin of error for a sample mean

When a research question asks you to estimate a parameter based on a numerical variable (for example, "What's the average age of teachers?"), the statistic used to help estimate the results is the average of all the responses provided by people in the sample. This is known as the *sample mean* (or average — see Chapter 5). And just like for sample proportions, you need to report a MOE for sample means.

The general formula for margin of error for the sample mean (assuming a certain condition is met) is $z^* \frac{\sigma}{\sqrt{n}}$, where σ is the population standard deviation, n is the sample size, and z^* is the appropriate z^*-value for your desired level of confidence (which you can find in Table 12-1).

Here are the steps for calculating the margin of error for a sample mean:

1. **Find the population standard deviation, σ, and the sample size, n.**

    The population standard deviation will be given in the problem.

2. **Divide the population standard deviation by the square root of the sample size.**

    $\frac{\sigma}{\sqrt{n}}$ gives you the standard error.

3. **Multiply by the appropriate z^*-value (refer to Table 12-1).**

    For example, the z^*-value is 1.96 if you want to be about 95% confident.

The condition you need to meet in order to use a z^*-value in the margin of error formula for a sample mean is either: 1) The original population has a normal distribution to start with, or 2) The sample size is large enough so the normal distribution can be used (that is, the Central Limit Theorem kicks in; see Chapter 11). In general, the sample size, n, should be above about 30 for the Central Limit Theorem. Now, if it's 29, don't panic — 30 is not a magic number, it's just a general rule of thumb. (The population standard deviation must be known either way.)

Suppose you're the manager of an ice cream shop, and you're training new employees to be able to fill the large-size cones with the proper amount of ice cream (10 ounces each). You want to estimate the average weight of the cones they make over a one-day period, including a margin of error. Instead of weighing every single cone made, you ask each of your new employees to randomly spot check the weights of a random sample of the large cones they make and record those weights on a notepad. For $n = 50$ cones sampled, the sample mean was found to be 10.3 ounces. Suppose the population standard deviation of $\sigma = 0.6$ ounces is known.

What's the margin of error? (Assume you want a 95% level of confidence.) It's calculated this way:

$$z^* \frac{\sigma}{\sqrt{n}} = 1.96 \frac{0.6}{\sqrt{50}} = (1.96)(0.0849) = 0.17$$

So to report these results, you say that based on the sample of 50 cones, you estimate that the average weight of all large cones made by the new employees over a one-day period is 10.3 ounces, with a margin of error of plus or minus 0.17 ounces. In other words, the range of likely values for the average weight of all large cones made for the day is estimated (with 95% confidence) to be between $10.30 - 0.17 = 10.13$ ounces and $10.30 + 0.17 = 10.47$ ounces. The new employees appear to be giving out too much ice cream (but I have a feeling the customers aren't offended).

Notice in the ice-cream-cone example, the units are ounces, not percentages! When working with and reporting results about data, always remember what the units are. Also, be sure that statistics are reported with their correct units of measure, and if they're not, ask what the units are.

In cases where n is too small (in general, less than 30) for the Central Limit Theorem to be used, but you still think the data came from a normal distribution, you can use a t^*-value instead of a z^*-value in your formulas. A t^*-value is one that comes from a t-distribution with $n - 1$ degrees of freedom. (Chapter 10 gives you all the in-depth details on the t-distribution.) In fact, many statisticians go ahead and use t^*-values instead of z^*-values consistently, because if the sample size is large, t^*-values and z^*-values are approximately equal anyway. In addition, for cases where you don't know the population standard deviation, σ, you can substitute it with s, the sample standard deviation; from there you use a t^*-value instead of a z^*-value in your formulas as well.

Being confident you're right

If you want to be *more* than 95% confident about your results, you need to add and subtract more than 1.96 standard errors (see Table 12-1). For example, to be 99% confident, you add and subtract 2.58 standard errors to obtain your margin of error. More confidence means a larger margin of error, though (assuming the sample size stays the same); so you have to ask yourself if it's worth it. When going from 95% to 99% confidence, the z^*-value increases by

2.58 – 1.96 = 0.62 (see Table 12-1). Most people don't think adding and subtracting this much more of a MOE is worthwhile, just to be 4% more confident (99% versus 95%) in the results obtained.

You can never be completely certain that your sample results do reflect the population, even with the margin of error included. Even if you're 95% confident in your results, that actually means that if you repeat the sampling process over and over, 5% of the time the sample won't represent the population well, simply due to chance (not because of problems with the sampling process or anything else). In these cases, you would miss the mark. So all results need to be viewed with that in mind.

Determining the Impact of Sample Size

The two most important ideas regarding sample size and margin of error are the following:

- ✔ Sample size and margin of error have an inverse relationship.
- ✔ After a point, increasing *n* beyond what you already have gives you a diminished return.

This section illustrates both concepts.

Sample size and margin of error

The relationship between margin of error and sample size is simple: As the sample size increases, the margin of error decreases. This relationship is called an inverse because the two move in opposite directions. If you think about it, it makes sense that the more information you have, the more accurate your results are going to get (in other words, the smaller your margin of error will get). (That assumes, of course, that the data were collected and handled properly.)

In the previous section, you see that the impact of a larger confidence level is a larger MOE. But if you increase the sample size, you can offset the larger MOE and bring it down to a reasonable size! Find out more about this concept in Chapter 13.

Bigger isn't always (that much) better!

In the example of the poll involving the approval rating of the president (see the earlier section "Calculating margin of error for a sample proportion"), the results of a sample of only 1,000 people from well over 310,000,000 residents in the United States could get to within about 3% of what the whole population would have said, if they had all been asked.

Using the formula for margin of error for a sample proportion, you can look at how the margin of error changes dramatically for samples of different sizes. Suppose in the presidential approval poll that n was 500 instead of 1,000. (Recall that $\hat{p} = 0.52$ for this example.) Therefore the margin of error for 95% confidence

is $1.96\sqrt{\dfrac{(0.52)(0.48)}{500}} = (1.96)(0.0223) = 0.0438$, which is equivalent to 4.38%.

When $n = 1,000$ in the same example, the margin of error (for 95% confidence)

is $1.96\sqrt{\dfrac{(0.52)(0.48)}{1,000}} = (1.96)(0.0158) = 0.0310$, which is equal to 3.10%. If n

is increased to 1,500, the margin of error (with the same level of confidence)

becomes $1.96\sqrt{\dfrac{(0.52)(0.48)}{1,500}} = (1.96)(0.0129) = 0.0253$, or 2.53%. Finally, when

$n = 2,000$, the margin of error is $1.96\sqrt{\dfrac{(0.52)(0.48)}{2,000}} = (1.96)(0.0112) = 0.0219$, or 2.19%.

Looking at these different results, you can see that larger sample sizes decrease the MOE, but after a certain point, you have a diminished return. Each time you survey one more person, the cost of your survey increases, and going from a sample size of, say, 1,500 to a sample size of 2,000 decreases your margin of error by only 0.34% (one third of one percent!) — from 0.0253 to 0.0219. The extra cost and trouble to get that small decrease in the MOE may not be worthwhile. Bigger isn't always that much better!

But what may really surprise you is that bigger can actually be worse! I explain this surprising fact in the following section.

Keeping margin of error in perspective

The margin of error is a measure of how close you expect your sample results to represent the entire population being studied. (Or at least it gives an upper limit for the amount of error you should have.) Because you're basing your conclusions about the population on your one sample, you have to account for how much those sample results could vary just due to chance.

Another view of margin of error is that it represents the maximum expected distance between the sample results and the actual population results (if you'd been able to obtain them through a census). Of course if you had the absolute truth about the population, you wouldn't be trying to do a survey, would you?

Just as important as knowing what the margin of error measures is realizing what the margin of error does *not* measure. The margin of error does not measure anything other than chance variation. That is, it doesn't measure any bias or errors that happen during the selection of the participants, the preparation or conduct of the survey, the data collection and entry process, or the analysis of the data and the drawing of the final conclusions.

A good slogan to remember when examining statistical results is "garbage in equals garbage out." No matter how nice and scientific the margin of error may look, remember that the formula that was used to calculate it doesn't have any idea of the quality of the data that the margin of error is based on. If the sample proportion or sample mean was based on a *biased sample* (one that favored certain people over others), a bad design, bad data-collection procedures, biased questions, or systematic errors in recording, then calculating the margin of error is pointless because it won't mean a thing.

For example, 50,000 people surveyed sounds great, but if they were all visitors to a certain Web site, the margin of error for this result is bogus because the calculation is all based on biased results! In fact, many extremely large samples are the result of biased sampling procedures. Of course, some people go ahead and report them anyway, so you have to find out what went into the formula: good information or garbage? If it turns out to be garbage, you know what to do about the margin of error. Ignore it. (For more information on errors that can take place during a survey or experiment, see Chapters 16 and 17, respectively.)

The Gallup Organization addresses the issue of what margin of error does and doesn't measure in a disclaimer that it uses to report its survey results. Gallup tells you that besides sampling error, surveys can have additional errors or bias due to question wording and some of the logistical issues involved in conducting surveys (such as missing data due to phone numbers that are no longer current).

This means that even with the best of intentions and the most meticulous attention to details and process control, stuff happens. Nothing is ever perfect. But what you need to know is that the margin of error can't measure the extent of those other types of errors. And if a highly credible polling organization like Gallup admits to possible bias, imagine what's really going on with other people's studies that aren't nearly as well designed or conducted.

Chapter 13

Confidence Intervals: Making Your Best Guesstimate

*M*ost statistics are used to estimate some characteristic about a population of interest, such as average household income, the percentage of people who buy birthday gifts online, or the average amount of ice cream consumed in the United States every year (and the resulting average weight gain — nah!). Such characteristics of a population are called *parameters*. Typically, people want to estimate (take a good guess at) the value of a parameter by taking a sample from the population and using statistics from the sample that will give them a good estimate. The question is: How do you define "good estimate"?

As long as the process is done correctly (and in the media, it often isn't!), an estimate can often get very close to the parameter. This chapter gives you an overview of confidence intervals (the type of estimates used and recommended by statisticians); why they should be used (as opposed to just a one-number estimate); how to set up, calculate, and interpret the most commonly used confidence intervals; and how to spot misleading estimates.

Not All Estimates Are Created Equal

Read any magazine or newspaper or listen to any newscast, and you hear a number of statistics, many of which are estimates of some quantity or another. You may wonder how they came up with those statistics. In some cases, the numbers are well researched; in other cases, they're just a shot in the dark. Here are some examples of estimates that I came across in one single issue of a leading business magazine. They come from a variety of sources:

- Even though some jobs are harder to get these days, some areas are really looking for recruits: Over the next eight years, 13,000 nurse anesthetists will be needed. Pay starts from $80,000 to $95,000.
- The average number of bats used by a major league baseball player per season is 90.
- The Lamborghini Murcielago can go from 0 to 60 mph in 3.7 seconds with a top speed of near 205 miles per hour.

Some of these estimates are easier to obtain than others. Here are some observations I was able to make about those estimates:

- How do you estimate how many nurse anesthetists are needed over the next eight years? You can start by looking at how many will be retiring in that time; but that won't account for growth. A prediction of the need in the next year or two would be close, but eight years into the future is much harder to do.
- The average number of bats used per major league baseball player in a season could be found by surveying the players themselves, the people who take care of their equipment, or the bat companies that supply the bats.
- Determining car speed is more difficult but could be conducted as a test with a stopwatch. And they should find the average speed of many different cars (not just one) of the same make and model, under the same driving conditions each time.

 Not all statistics are created equal. To determine whether a statistic is reliable and credible, don't just take it at face value. Think about whether it makes sense and how you would go about formulating an estimate. If the statistic is really important to you, find out what process was used to come up with it. (Chapter 16 handles all the elements involving surveys, and Chapter 17 gives you the lowdown on experiments.)

Linking a Statistic to a Parameter

A *parameter* is a single number that describes a population, such as the median household income for all households in the U.S. A *statistic* is a single number that describes a sample, such as the median household income of a sample of, say, 1,200 households. You typically don't know the values of parameters of populations, so you take samples and use statistics to give your best estimates.

Suppose you want to know the percentage of vehicles in the U.S. that are pickup trucks (that's the parameter, in this case). You can't look at every single vehicle, so you take a random sample of 1,000 vehicles over a range of highways at different times of the day. You find that 7% of the vehicles in your sample are pickup trucks. Now, you don't want to say that *exactly* 7% of all vehicles on U.S. roads are pickup trucks, because you know this is only based on the 1,000 vehicles you sampled. Though you hope 7% is close to the true percentage, you can't be sure because you based your results on a sample of vehicles, not on all the vehicles in the U.S.

So what to do? You take your sample result and add and subtract some number to indicate that you are giving a range of possible values for the population parameter, rather than just assuming the sample statistic equals the population parameter (which would not be good, although it's done in the media all the time). This number that is added to and subtracted from a statistic is called the *margin of error* (*MOE*). This plus or minus (denoted by ±) that's added to any estimate helps put the results into perspective. When you know the margin of error, you have an idea of how much the sample results could change if you took another sample.

The word *error* in *margin of error* doesn't mean a mistake was made or the quality of the data was bad. It just means the results from a sample are not exactly equal to what you would have gotten if you had used the entire population. This gap measures error due to random chance, the luck of the draw — not due to bias. (That's why minimizing bias is so important when you select your sample and collect your data; see Chapters 16 and 17.)

Getting with the Jargon

A statistic plus or minus a margin of error is called a *confidence interval:*

- ✔ The word *interval* is used because your result becomes an interval. For example, say the percentage of kids who like baseball is 40%, plus or minus 3.5%. That means the percentage of kids who like baseball is somewhere between 40% − 3.5% = 36.5% and 40% + 3.5% = 43.5%. The lower end of the interval is your statistic minus the margin of error, and the upper end is your statistic plus the margin of error.

- ✔ With all confidence intervals, you have a certain amount of confidence in being correct (guessing the parameter) with your sample in the long run. Expressed as a percent, the amount of confidence is called the *confidence level.*

You can find formulas and examples for the most commonly used confidence intervals later in this chapter.

Following are the general steps for estimating a parameter with a confidence interval. Details on Steps 1 and 4–6 are included throughout the remainder of this chapter. Steps 2 and 3 involve sampling and data collection, which are detailed in Chapter 16 (sampling and survey data collection) and Chapter 17 (data collection from experiments).

1. **Choose your confidence level and your sample size.**

2. **Select a random sample of individuals from the population.**

3. **Collect reliable and relevant data from the individuals in the sample.**

4. **Summarize the data into a statistic, such as a mean or proportion.**

5. **Calculate the margin of error.**

6. **Take the statistic plus or minus the margin of error to get your final estimate of the parameter.**

   This step calculates the *confidence interval* for that parameter.

Interpreting Results with Confidence

Suppose you, a research biologist, are trying to catch a fish using a hand net, and the size of your net represents the margin of error of a confidence interval. Now say your confidence level is 95%. What does this really mean? It means that if you scoop this particular net into the water over and over again, you'll catch a fish 95% of the time. Catching a fish here means your confidence interval was correct and contains the true parameter (in this case the parameter is represented by the fish itself).

But does this mean that on any given try you have a 95% chance of catching a fish after the fact? No. Is this confusing? It certainly is. Here's the scoop (no pun intended): On a single try, say you close your eyes before you scoop your net into the water. At this point, your chances of catching a fish are 95%. But then go ahead and scoop your net through the water with your eyes still closed. *After* that's done, however, you open your eyes and see one of only two possible outcomes; you either caught a fish or you didn't; probability isn't involved anymore.

Likewise, *after* data have been collected, and the confidence interval has been calculated, you either captured the true population parameter or you didn't. So you're not saying you're 95% confident that the parameter is in your particular interval. What you are 95% confident about is the process by which random samples are selected and confidence intervals are created. (That is, 95% of the time in the long run, you'll catch a fish.)

You know that this process will result in intervals that capture the population mean 95% of the time. The other 5% of the time, the data collected in the sample just by random chance has abnormally high or low values in it and doesn't represent the population. This 5% measures errors due to random chance only and doesn't include bias.

The margin of error is meaningless if the data that went into the study were biased and/or unreliable. However, you can't tell that by looking at anyone's statistical results. My best advice is to look at how the data were collected before accepting a reported margin of error as the truth (see Chapters 16 and 17 for details on data collection issues). That means asking questions before you believe a study.

Zooming In on Width

The *width* of your confidence interval is two times the margin of error. For example, suppose the margin of error is ± 5%. A confidence interval of 7%, plus or minus 5%, goes from 7% – 5% = 2%, all the way up to 7% + 5% = 12%. So the confidence interval has a width of 12% – 2% = 10%. A simpler way to calculate this is to say that the width of the confidence interval is two times the margin of error. In this case, the width of the confidence interval is 2 * 5% = 10%.

The width of a confidence interval is the distance from the lower end of the interval (statistic minus margin of error) to the upper end of the interval (statistic plus margin of error). You can always calculate the width of a confidence interval quickly by taking two times the margin of error.

The ultimate goal when making an estimate using a confidence interval is to have a narrow width, because that means you're zooming in on what the parameter is. Having to add and subtract a large margin of error only makes your result much less accurate.

So, if a small margin of error is good, is smaller even better? Not always. A narrow confidence interval is a good thing — to a point. To get an extremely narrow confidence interval, you have to conduct a much larger — and expensive — study, so a point comes where the increase in price doesn't justify the marginal difference in accuracy. Most people are pretty comfortable with a margin of error of 2% to 3% when the estimate itself is a percentage (like the percentage of women, Republicans, or smokers).

How do you go about ensuring that your confidence interval will be narrow enough? You certainly want to think about this issue before collecting your data; after the data are collected, the width of the confidence interval is set.

Three factors affect the width of a confidence interval:

- ✔ Confidence level
- ✔ Sample size
- ✔ Amount of variability in the population

Each of these three factors plays an important role in influencing the width of a confidence interval. In the following sections, you explore details of each element and how they affect width.

Choosing a Confidence Level

Every confidence interval (and every margin of error, for that matter) has a percentage associated with it that represents how confident you are that the results will capture the true population parameter, depending on the luck of the draw with your random sample. This percentage is called a *confidence level*.

A confidence level helps you account for the other possible sample results you could have gotten, when you're making an estimate of a parameter using the data from only one sample. If you want to account for 95% of the other possible results, your confidence level would be 95%.

What level of confidence is typically used by researchers? I've seen confidence levels ranging from 80% to 99%. The most common confidence level is 95%. In fact, statisticians have a saying that goes, "Why do statisticians like their jobs? Because they have to be correct only 95% of the time." (Sort of catchy, isn't it? And let's see weather forecasters beat that.)

Variability in sample results is measured in terms of number of standard errors. A *standard error* is similar to the standard deviation of a data set, only a standard error applies to sample means or sample percentages that you could have gotten if different samples were taken. (See Chapter 11 for information on standard errors.)

Standard errors are the building blocks of confidence intervals. A confidence interval is a statistic plus or minus a margin of error, and the margin of error is the number of standard errors you need to get the confidence level you want.

Every confidence level has a corresponding number of standard errors that have to be added or subtracted. This number of standard errors is a called a *critical value*. In a situation where you use a Z-distribution to find the number of standard errors (as described later in this chapter), you call the critical value the *z*-value* (pronounced *z-star value*). See Table 13-1 for a list of z*-values for some of the most common confidence levels.

As the confidence level increases, the number of standard errors increases, so the margin of error increases.

Table 13-1	*z**-values for Various Confidence Levels
Confidence Level	*z*-value*
80%	1.28
90%	1.645 (by convention)
95%	1.96
98%	2.33
99%	2.58

If you want to be more than 95% confident about your results, you need to add and subtract more than about two standard errors. For example, to be 99% confident, you would add and subtract about two and a half standard errors to obtain your margin of error (2.58 to be exact). The higher the confidence level, the larger the z^*-value, the larger the margin of error, and the wider the confidence interval (assuming everything else stays the same). You have to pay a certain price for more confidence.

Note that I said "assuming everything else stays the same." You can offset an increase in the margin of error by increasing the sample size. See the following section for more on this.

Factoring In the Sample Size

The relationship between margin of error and sample size is simple: As the sample size increases, the margin of error decreases, and the confidence interval gets narrower. This relationship confirms what you hope is true: The more information (data) you have, the more accurate your results are going to be. (That, of course, assumes that the information is good, credible information. See Chapter 3 for how statistics can go wrong.)

The margin of error formulas for the confidence intervals in this chapter all involve the sample size (n) in the denominator. For example, the formula for margin of error for the sample mean, $\pm z^* \dfrac{\sigma}{\sqrt{n}}$ (which you'll see in great detail later in this chapter), has an n in the denominator of a fraction (this is the case for most margin of error formulas). As n increases, the denominator of this fraction increases, which makes the overall fraction get smaller. That makes the margin of error smaller and results in a narrower confidence interval.

When you need a high level of confidence, you have to increase the z^*-value and, hence, margin of error, resulting in a wider confidence interval, which isn't good. (See the previous section.) But you can offset this wider confidence interval by increasing the sample size and bringing the margin of error back down, thus narrowing the confidence interval.

The increase in sample size allows you to still have the confidence level you want, but also ensures that the width of your confidence interval will be small (which is what you ultimately want). You can even determine the sample size you need before you start a study: If you know the margin of error you want to get, you can set your sample size accordingly. (See the later section "Figuring Out What Sample Size You Need" for more.)

When your statistic is going to be a percentage (such as the percentage of people who prefer to wear sandals during summer), a rough way to figure margin of error for a 95% confidence interval is to take 1 divided by the square root of n (the sample size). You can try different values of n and you can see how the margin of error is affected. For example, a survey of 100 people from a large population will have a margin of error of about $\frac{1}{\sqrt{100}} = 0.10$ or plus or minus 10% (meaning the width of the confidence interval is 20%, which is pretty large).

However, if you survey 1,000 people, your margin of error decreases dramatically, to plus or minus about 3%; the width now becomes only 6%. A survey of 2,500 people results in a margin of error of plus or minus 2% (so the width is down to 4%). That's quite a small sample size to get so accurate, when you think about how large the population is (the U.S. population, for example, is over 310 million!).

Keep in mind, however, you don't want to go *too* high with your sample size, because a point comes where you have a diminished return. For example, moving from a sample size of 2,500 to 5,000 narrows the width of the confidence interval to about $2 * 1.4 = 2.8\%$, down from 4%. Each time you survey one more person, the cost of your survey increases, so adding another 2,500 people to the survey just to narrow the interval by little more than 1% may not be worthwhile.

The first step in any data analysis problem (and when critiquing another person's results) is to make sure you have good data. Statistical results are only as good as the data that went into them, so real accuracy depends on the quality of the data as well as on the sample size. A large sample size that has a great deal of bias (see Chapter 16) may appear to have a narrow confidence interval — but means nothing. That's like competing in an archery match and shooting your arrows consistently, but finding out that the whole time you're shooting at the next person's target; that's how far off you are. With the field of statistics, though, you can't accurately measure bias; you can only try to minimize it by designing good samples and studies (see Chapters 16 and 17).

Counting On Population Variability

One of the factors influencing variability in sample results is the fact that the population itself contains variability. For example, in a population of houses in a fairly large city like Columbus, Ohio, you see a great deal of variety in not only the types of houses, but also the sizes and the prices. And the variability in prices of houses in Columbus should be more than the variability in prices of houses in a selected housing development in Columbus.

That means if you take a sample of houses from the entire city of Columbus and find the average price, the margin of error should be larger than if you take a sample from that single housing development in Columbus, even if you have the same confidence level and the same sample size.

Why? Because the houses in the entire city have more variability in price, and your sample average would change more from sample to sample than it would if you took the sample only from that single housing development, where the prices tend to be very similar because houses tend to be comparable in a single housing development. So you need to sample more houses if you're sampling from the entire city of Columbus in order to have the same amount of accuracy that you would get from that single housing development.

The standard deviation of the population is denoted σ. Notice that σ appears in the numerator of the standard error in the formula for margin of error for the sample mean: $\pm z^* \dfrac{\sigma}{\sqrt{n}}$.

Therefore, as the standard deviation (the numerator) increases, the standard error (the entire fraction) also increases. This results in a larger margin of error and a wider confidence interval. (Refer to Chapter 11 for more info on the standard error.)

More variability in the original population increases the margin of error, making the confidence interval wider. This increase can be offset by increasing the sample size.

Calculating a Confidence Interval for a Population Mean

When the characteristic that's being measured (such as income, IQ, price, height, quantity, or weight) is *numerical,* most people want to estimate the mean (average) value for the population. You estimate the population mean, μ, by using a sample mean, $\bar{x}$, plus or minus a margin of error. The result is called a *confidence interval for the population mean, μ.* Its formula depends on whether certain conditions are met. I split the conditions into two cases, illustrated in the following sections.

Case 1: Population standard deviation is known

In Case 1, the population standard deviation is known. The formula for a confidence interval (CI) for a population mean in this case is $\bar{x} \pm z^* \frac{\sigma}{\sqrt{n}}$, where $\bar{x}$ is the sample mean, σ is the population standard deviation, n is the sample size, and z^* represents the appropriate z^*-value from the standard normal distribution for your desired confidence level. (Refer to Table 13-1 for values of z^* for the given confidence levels.)

In this case, the data either have to come from a normal distribution, or if not, then n has to be large enough (at least 30 or so) for the Central Limit Theorem to kick in (see Chapter 11), allowing you to use z^*-values in the formula.

To calculate a CI for the population mean (average), under the conditions for Case 1, do the following:

1. **Determine the confidence level and find the appropriate z^*-value.**

   Refer to Table 13-1.

2. **Find the sample mean ($\bar{x}$) for the sample size (n).**

   Note: The population standard deviation is assumed to be a known value, σ.

3. **Multiply z^* times σ and divide that by the square root of n.**

   This calculation gives you the margin of error.

4. **Take $\bar{x}$ plus or minus the margin of error to obtain the CI.**

   The lower end of the CI is $\bar{x}$ minus the margin of error, whereas the upper end of the CI is $\bar{x}$ plus the margin of error.

For example, suppose you work for the Department of Natural Resources and you want to estimate, with 95% confidence, the mean (average) length of walleye fingerlings in a fish hatchery pond.

1. Because you want a 95% confidence interval, your z^*-value is 1.96.

2. Suppose you take a random sample of 100 fingerlings and determine that the average length is 7.5 inches; assume the population standard deviation is 2.3 inches. This means $\bar{x} = 7.5$, $\sigma = 2.3$, and $n = 100$.

3. Multiply 1.96 times 2.3 divided by the square root of 100 (which is 10). The margin of error is, therefore, $\pm 1.96 * (2.3 \div 10) = 1.96 * 0.23 = 0.45$ inches.

4. Your 95% confidence interval for the mean length of walleye fingerlings in this fish hatchery pond is 7.5 inches ± 0.45 inches. (The lower end of the interval is $7.5 - 0.45 = 7.05$ inches; the upper end is $7.5 + 0.45 = 7.95$ inches.)

After you calculate a confidence interval, make sure you always interpret it in words a non-statistician would understand. That is, talk about the results in terms of what the person in the problem is trying to find out — statisticians call this interpreting the results "in the context of the problem." In this example you can say: "With 95% confidence, the average length of walleye fingerlings in this entire fish hatchery pond is between 7.05 and 7.95 inches, based on my sample data." (Always be sure to include appropriate units.)

Case 2: Population standard deviation is unknown and/or n is small

In many situations, you don't know σ, so you estimate it with the sample standard deviation, $s;$ and/or the sample size is small (less than 30), and you can't be sure your data came from a normal distribution. (In the latter case, the Central Limit Theorem can't be used; see Chapter 11.) In either situation, you can't use a z^*-value from the standard normal (Z-) distribution as your critical value anymore; you have to use a larger critical value than that, because of not knowing what σ is and/or having less data.

The formula for a confidence interval for one population mean in Case 2 is $\bar{x} \pm t^*_{n-1} \dfrac{s}{\sqrt{n}}$, where t^*_{n-1} is the critical t^*-value from the t-distribution with $n - 1$ degrees of freedom (where n is the sample size). The t^*-values for common confidence levels are found using the last row of the t-table (in the appendix). Chapter 10 gives you the full details on the t-distribution and how to use the t-table.

The t-distribution has a similar shape to the Z-distribution except it's flatter and more spread out. For small values of n and a specific confidence level, the critical values on the t-distribution are larger than on the Z-distribution, so when you use the critical values from the t-distribution, the margin of error for your confidence interval will be wider. As the values of n get larger, the t^*-values are closer to z^*-values. (Chapter 10 gives you the full details on the t-distribution and its relationships to the Z-distribution.)

In the fish hatchery example from Case 1, suppose your sample size was 10 instead of 100, and everything else was the same. The t^*-value in this case comes from a t-distribution with $10 - 1 = 9$ degrees of freedom. This t^*-value is found by looking at the t-table (in the appendix). Look in the last row where the confidence levels are located, and find the confidence level of 95%; this marks the column you need. Then find the row corresponding to $df = 9$. Intersect the row and column, and you find $t^* = 2.262$. This is the t^*-value for a 95% confidence interval for the mean with a sample size of 10. (Notice this is larger than the z^*-value of 1.96 found in Table 13-1.) Calculating the confidence interval, you get $7.5 \pm 2.262 \dfrac{2.3}{\sqrt{10}} = 7.50 \pm 1.645$, or 5.86 to 9.15 inches. (Chapter 10 gives you the full details on the t-distribution and how to use the t-table.)

Notice this confidence interval is wider than the one found when $n = 100$. In addition to having a larger critical value (t^* versus z^*), the sample size is much smaller, which increases the margin of error, because n is in its denominator.

In a case where you need to use s because you don't know σ, the confidence interval will be wider as well. It is also often the case that σ is unknown and the sample size is small, in which case the confidence interval is also wider.

Figuring Out What Sample Size You Need

The margin of error of a confidence interval is affected by size (see the earlier section "Factoring In the Sample Size"); as size increases, margin of error decreases. Looking at this the other way around, if you want a smaller margin of error (and doesn't everyone?), you need a larger sample size. Suppose you are getting ready to do your own survey to estimate a population mean; wouldn't it be nice to see ahead of time what sample size you need to get the margin of error you want? Thinking ahead will save you money and time and it will give you results you can live with in terms of the margin of error — you won't have any surprises later.

The formula for the sample size required to get a desired margin of error (MOE) when you are doing a confidence interval for μ is $n \geq \left(\dfrac{z^* \sigma}{MOE} \right)^2$; always round up the sample size no matter what decimal value you get. (For example, if your calculations give you 126.2 people, you can't just have 0.2 of a person — you need the whole person, so include him by rounding up to 127.)

In this formula, MOE is the number representing the margin of error you want, and z^* is the z^*-value corresponding to your desired confidence level (from Table 13-1; most people use 1.96 for a 95% confidence interval). If the population standard deviation, σ, is unknown, you can put in a worst-case scenario guess for it or run a pilot study (a small trial study) ahead of time, find the standard deviation of the sample data (s), and use that number. This can be risky if the sample size is very small because it's less likely to reflect the whole population; try to get the largest trial study that you can, and/or make a conservative estimate for σ.

Often a small trial study is worth the time and effort. Not only will you get an estimate of σ to help you determine a good sample size, but you may also learn about possible problems in your data collection.

I only include one formula for calculating sample size in this chapter: the one that pertains to a confidence interval for a population mean. (You can, however, use the quick and dirty formula in the earlier section "Factoring in the Sample Size" for handling proportions.)

Here's an example where you need to calculate n to estimate a population mean. Suppose you want to estimate the average number of songs college students store on their portable devices. You want the margin of error to be *no more than* plus or minus 20 songs. You want a 95% confidence interval. How many students should you sample?

Because you want a 95% CI, z^* is 1.96 (found in Table 13-1); you know your desired MOE is 20. Now you need a number for the population standard deviation, σ. This number is not known, so you do a pilot study of 35 students and find the standard deviation (s) for the sample is 148 songs — use this number as a substitute for σ. Using the sample size formula, you calculate the sample size you need is $n \geq \left(\dfrac{1.96(148)}{20} \right)^2 = (14.504)^2 = 210.37$, which you round *up* to 211 students (you always round up when calculating n). So you need to take a random sample of *at least* 211 college students in order to have a margin of error in the number of stored songs of *no more than* 20. That's why you see a greater-than-or-equal-to sign in the formula here.

You always round up to the nearest integer when calculating sample size, no matter what the decimal value of your result is (for example, 0.37). That's because you want the margin of error to be *no more than* what you stated. If you round down when the decimal value is under .50 (as you normally do in other math calculations), your MOE will be a little larger than you wanted.

If you are wondering where this formula for sample size came from, it's actually created with just a little math gymnastics. Take the margin of error formula (which contains n), fill in the remaining variables in the formula with numbers you glean from the problem, set it equal to the desired MOE, and solve for n.

Determining the Confidence Interval for One Population Proportion

When a characteristic being measured is categorical — for example, opinion on an issue (support, oppose, or are neutral), gender, political party, or type of behavior (do/don't wear a seatbelt while driving) — most people want to estimate the proportion (or percentage) of people in the population that fall into a certain category of interest. For example, consider the percentage of people in favor of a four-day work week, the percentage of Republicans who voted in the last election, or the proportion of drivers who don't wear seat belts. In each of these cases, the object is to estimate a population proportion, p, using a sample proportion, $\hat{p}$, plus or minus a margin of error. The result is called a *confidence interval for the population proportion, p.*

The formula for a CI for a population proportion is $\hat{p} \pm z^* \sqrt{\dfrac{\hat{p}(1-\hat{p})}{n}}$, where $\hat{p}$ is the sample proportion, n is the sample size, and z^* is the appropriate value from the standard normal distribution for your desired confidence level. Refer to Table 13-1 for values of z^* for certain confidence levels.

To calculate a CI for the population proportion:

1. **Determine the confidence level and find the appropriate z^*-value.**

   Refer to Table 13-1 for z^*-values.

2. **Find the sample proportion, $\hat{p}$, by dividing the number of people in the sample having the characteristic of interest by the sample size (n).**

   Note: This result should be a decimal value between 0 and 1.

3. **Multiply $\hat{p}(1 - \hat{p})$ and then divide that amount by n.**

4. **Take the square root of the result from Step 3.**

5. **Multiply your answer by z^*.**

   This step gives you the margin of error.

6. **Take $\hat{p}$ plus or minus the margin of error to obtain the CI; the lower end of the CI is $\hat{p}$ minus the margin of error, and the upper end of the CI is $\hat{p}$ plus the margin of error.**

The formula shown in the preceding example for a CI for *p* is used under the condition that the sample size is large enough for the Central Limit Theorem to kick in and allow us to use a z^*-value (see Chapter 11), which happens in cases when you are estimating proportions based on large scale surveys (see Chapter 9). For small sample sizes, confidence intervals for the proportion are typically beyond the scope of an intro statistics course.

For example, suppose you want to estimate the percentage of the time you're expected to get a red light at a certain intersection.

1. Because you want a 95% confidence interval, your z^*-value is 1.96.

2. You take a random sample of 100 different trips through this intersection and find that you hit a red light 53 times, so $\hat{p} = 53 \div 100 = 0.53$.

3. Find $\hat{p}(1-\hat{p}) = 0.53 * (1 - 0.53) = 0.2491 \div 100 = 0.002491$.

4. Take the square root to get 0.0499.

   The margin of error is, therefore, plus or minus $1.96 * (0.0499) = 0.0978$, or 9.78%.

5. Your 95% confidence interval for the percentage of times you will ever hit a red light at that particular intersection is 0.53 (or 53%), plus or minus 0.0978 (rounded to 0.10 or 10%). (The lower end of the interval is 0.53 − 0.10 = 0.43 or 43%; the upper end is 0.53 + 0.10 = 0.63 or 63%.)

   To interpret these results within the context of the problem, you can say that with 95% confidence the percentage of the times you should expect to hit a red light at this intersection is somewhere between 43% and 63%, based on your sample.

While performing any calculations involving sample percentages, use the decimal form. After the calculations are finished, convert to percentages by multiplying by 100. To avoid round-off error, keep at least 2 decimal places throughout.

Creating a Confidence Interval for the Difference of Two Means

The goal of many surveys and studies is to compare two populations, such as men versus women, low versus high income families, and Republicans versus Democrats. When the characteristic being compared is numerical (for example, height, weight, or income), the object of interest is the amount of difference in the means (averages) for the two populations.

For example, you may want to compare the difference in average age of Republicans versus Democrats, or the difference in average incomes of men versus women. You estimate the difference between two population means, $\mu_1 - \mu_2$, by taking a sample from each population (say, sample 1 and sample 2) and using the difference of the two sample means $\bar{x}_1 - \bar{x}_2$, plus or minus a margin of error. The result is a *confidence interval for the difference of two population means, $\mu_1 - \mu_2$*. The formula for the CI is different depending on certain conditions, as seen in the following sections; I call them Case 1 and Case 2.

Case 1: Population standard deviations are known

Case 1 assumes that both of the population standard deviations are known. The formula for a CI for the difference between two population means (averages) is

$\bar{x}_1 - \bar{x}_2 \pm z^* \sqrt{\dfrac{\sigma_1^2}{n_1} + \dfrac{\sigma_2^2}{n_2}}$, where $\bar{x}_1$ and n_1 are the mean and size of the first sample,

and the first population's standard deviation, σ_1, is given (known); $\bar{x}_2$ and n_2 are the mean and size of the second sample, and the second population's standard deviation, σ_2, is given (known). Here z^* is the appropriate value from the standard normal distribution for your desired confidence level. (Refer to Table 13-1 for values of z^* for certain confidence levels.)

To calculate a CI for the difference between two population means, do the following:

1. **Determine the confidence level and find the appropriate z^*-value.**

   Refer to Table 13-1.

2. **Identify $\bar{x}_1$, n_1, and σ_1; find $\bar{x}_2$, n_2, and σ_2.**

3. **Find the difference, $(\bar{x}_1 - \bar{x}_2)$, between the sample means.**

4. **Square σ_1 and divide it by n_1; square σ_2 and divide it by n_2. Add the results together and take the square root.**

5. **Multiply your answer from Step 4 by z^*.**

   This answer is the margin of error.

6. **Take $\bar{x}_1 - \bar{x}_2$ plus or minus the margin of error to obtain the CI.**

   The lower end of the CI is $\bar{x}_1 - \bar{x}_2$ *minus* the margin of error, whereas the upper end of the CI is $\bar{x}_1 - \bar{x}_2$ *plus* the margin of error.

Suppose you want to estimate with 95% confidence the difference between the mean (average) length of the cobs of two varieties of sweet corn (allowing them to grow the same number of days under the same conditions). Call the two varieties Corn-e-stats and Stats-o-sweet. Assume by prior research that the population standard deviations for Corn-e-stats and Stats-o-sweet are 0.35 inches and 0.45 inches, respectively.

1. Because you want a 95% confidence interval, your z^* is 1.96.

2. Suppose your random sample of 100 cobs of the Corn-e-stats variety averages 8.5 inches, and your random sample of 110 cobs of Stats-o-sweet averages 7.5 inches. So the information you have is: $\bar{x}_1 = 8.5$, $\sigma_1 = 0.35$, $n_1 = 100$, $\bar{x}_2 = 7.5$, $\sigma_2 = 0.45$, and $n_2 = 110$.

3. The difference between the sample means, $\bar{x}_1 - \bar{x}_2$, from Step 3, is $8.5 - 7.5 = +1$ inch. This means the average for Corn-e-stats minus the average for Stats-o-sweet is positive, making Corn-e-stats the larger of the two varieties, in terms of this sample. Is that difference enough to generalize to the entire population, though? That's what this confidence interval is going to help you decide.

4. Square σ_1 (0.35) to get 0.1225; divide by 100 to get 0.0012. Square σ_2 (0.45) and divide by 110 to get $0.2025 \div 110 = 0.0018$. The sum is $0.0012 + 0.0018 = 0.0030$; the square root is 0.0554 inches (if no rounding was done).

5. Multiply 1.96 times 0.0554 to get 0.1085 inches, the margin of error.

6. Your 95% confidence interval for the difference between the average lengths for these two varieties of sweet corn is 1 inch, plus or minus 0.1085 inches. (The lower end of the interval is $1 - 0.1085 = 0.8915$ inches; the upper end is $1 + 0.1085 = 1.1085$ inches.) Notice all the values in this interval are positive. That means Corn-e-stats is estimated to be longer than Stats-o-sweet, based on your data.

   To interpret these results in the context of the problem, you can say with 95% confidence that the Corn-e-stats variety is longer, on average, than the Stats-o-sweet variety, by somewhere between 0.8915 and 1.1085 inches, based on your sample.

Notice that you could get a negative value for $\bar{x}_1 - \bar{x}_2$. For example, if you had switched the two varieties of corn, you would have gotten –1 for this difference. You would say that Stats-o-sweet averaged one inch shorter than Corn-e-stats in the sample (the same conclusion stated differently).

If you want to avoid negative values for the difference in sample means, always make the group with the larger sample mean your first group — all your differences will be positive (that's what I do).

Case 2: Population standard deviations are unknown and/or sample sizes are small

In many situations, you don't know σ_1 and σ_2, and you estimate them with the sample standard deviations, s_1, and s_2; and/or the sample sizes are small (less than 30) and you can't be sure whether your data came from a normal distribution.

A confidence interval for the difference in two population means under Case 2 is $(\bar{x}_1 - \bar{x}_2) \pm t^*_{n_1 + n_2 - 2} \sqrt{\dfrac{(n_1 - 1)s_1^2 + (n_2 - 1)s_2^2}{n_1 + n_2 - 2}}$, where t^* is the critical value from the t-distribution with $n_1 + n_2 - 2$ degrees of freedom; n_1 and n_2 are the two sample sizes, respectively; and s_1 and s_2 are the two sample standard deviations. This t^*-value is found on the t-table (in the appendix) by intersecting the row for $df = n_1 + n_2 - 2$ with the column for the confidence level you need, as indicated by looking at the last row of the table. (See Chapter 10.) Here we assume the population standard deviations are similar; if not, modify by using the standard error and degrees of freedom. See the end of the section on comparing two means in Chapter 15.

In the corn example from Case 1, suppose the mean cob lengths of the two brands of corn, Corn-e-stats (group 1) and Stats-o-sweet (group 2), are the same as they were before: $\bar{x}_1 = 8.5$ and $\bar{x}_2 = 7.5$ inches. But this time you don't know the population standard deviations, so you use the sample standard deviations instead — suppose they turn out to be $s_1 = 0.40$ and $s_2 = 0.50$ inches, respectively. Suppose the sample sizes, n_1 and n_2, are each only 15 in this case.

Calculating the CI, you first need to find the t^*-value on the t-distribution with $(15 + 15 - 2) = 28$ degrees of freedom. (Assume the confidence level is still 95%.) Using the t-table (in the appendix), look at the row for 28 degrees of freedom and the column representing a confidence level of 95% (see the labels on the last row of the table); intersect them and you see $t^*_{28} = 2.048$. Using the rest of the information you are given, the confidence interval for the difference in mean cob length for the two brands is $(8.5 - 7.5) \pm 2.048 \sqrt{\dfrac{(15-1)(0.4)^2 + (15-1)(0.5)^2}{15 + 15 - 2}}$ $= 1.0 \pm 2.048(0.45) = 1.00 \pm 0.9273$ inches.

That means a 95% CI for the difference in the mean cob lengths of these two brands of corn in this situation is (0.0727, 1.9273) inches, with Corn-e-stats coming out on top. (**Note:** This CI is wider than what was found in Case 1, as expected.)

Estimating the Difference of Two Proportions

When a characteristic, such as opinion on an issue (support/don't support), of the two groups being compared is *categorical,* people want to report on the differences between the two population proportions — for example, the difference between the proportion of women who support a four-day work week and the proportion of men who support a four-day work week. How do you do this?

You estimate the difference between two population proportions, $p_1 - p_2$, by taking a sample from each population and using the difference of the two sample proportions, $\hat{p}_1 - \hat{p}_2$, plus or minus a margin of error. The result is called a *confidence interval for the difference of two population proportions, $p_1 - p_2$.*

The formula for a CI for the difference between two population proportions is $\left(\hat{p}_1 - \hat{p}_2\right) \pm z^* \sqrt{\dfrac{\hat{p}_1\left(1-\hat{p}_1\right)}{n_1} + \dfrac{\hat{p}_2\left(1-\hat{p}_2\right)}{n_2}}$, where $\hat{p}_1$ and n_1 are the sample proportion and sample size of the first sample, and $\hat{p}_2$ and n_2 are the sample proportion and sample size of the second sample. z^* is the appropriate value from the standard normal distribution for your desired confidence level. (Refer to Table 13-1 for z^*-values.)

To calculate a CI for the difference between two population proportions, do the following:

1. **Determine the confidence level and find the appropriate z^*-value.**

   Refer to Table 13-1.

2. **Find the sample proportion $\hat{p}_1$ for the first sample by taking the total number from the first sample that are in the category of interest and dividing by the sample size, n_1. Similarly, find $\hat{p}_2$ for the second sample.**

3. **Take the difference between the sample proportions, $\hat{p}_1 - \hat{p}_2$.**

4. **Find $\hat{p}_1\left(1-\hat{p}_1\right)$ and divide that by n_1. Find $\hat{p}_2\left(1-\hat{p}_2\right)$ and divide that by n_2. Add these two results together and take the square root.**

5. **Multiply z^* times the result from Step 4.**

   This step gives you the margin of error.

6. **Take $\hat{p}_1 - \hat{p}_2$ plus or minus the margin of error from Step 5 to obtain the CI.**

   The lower end of the CI is $\hat{p}_1 - \hat{p}_2$ minus the margin of error, and the upper end of the CI is $\hat{p}_1 - \hat{p}_2$ plus the margin of error.

The formula shown here for a CI for $p_1 - p_2$ is used under the condition that both of the sample sizes are large enough for the Central Limit Theorem to kick in and allow us to use a z^*-value (see Chapter 11); this is true when you are estimating proportions using large scale surveys, for example. For small sample sizes, confidence intervals are beyond the scope of an intro statistics course.

Suppose you work for the Las Vegas Chamber of Commerce, and you want to estimate with 95% confidence the difference between the percentage of females who have ever gone to see an Elvis impersonator and the percentage of males who have ever gone to see an Elvis impersonator, in order to help determine how you should market your entertainment offerings.

1. Because you want a 95% confidence interval, your z^*-value is 1.96.

2. Suppose your random sample of 100 females includes 53 females who have seen an Elvis impersonator, so $\hat{p}_1$ is 53 ÷ 100 = 0.53. Suppose also that your random sample of 110 males includes 37 males who have ever seen an Elvis impersonator, so $\hat{p}_2$ is 37 ÷ 110 = 0.34.

3. The difference between these sample proportions (females – males) is 0.53 – 0.34 = 0.19.

4. Take 0.53 * (1 – 0.53) and divide that by 100 to get 0.2491 ÷ 100 = 0.0025. Then take 0.34 * (1 – 0.34) and divide that by 110 to get 0.2244 ÷ 110 = 0.0020. Add these two results to get 0.0025 + 0.0020 = 0.0045; the square root is 0.0671.

5. 1.96 * 0.0671 gives you 0.13, or 13%, which is the margin of error.

6. Your 95% confidence interval for the difference between the percentage of females who have seen an Elvis impersonator and the percentage of males who have seen an Elvis impersonator is 0.19 or 19% (which you got in Step 3), plus or minus 13%. The lower end of the interval is 0.19 – 0.13 = 0.06 or 6%; the upper end is 0.19 + 0.13 = 0.32 or 32%.

   To interpret these results within the context of the problem, you can say with 95% confidence that a higher percentage of females than males have seen an Elvis impersonator, and the difference in these percentages is somewhere between 6% and 32%, based on your sample.

Now I'm thinking there are some guys out there that wouldn't admit they'd ever seen an Elvis impersonator (although they've probably pretended to be one doing karaoke at some point). This may create some bias in the results. (The last time I was in Vegas, I believe I really saw Elvis; he was driving a van taxi to and from the airport. . . .)

Notice that you could get a negative value for $\hat{p}_1 - \hat{p}_2$. For example, if you had switched the males and females, you would have gotten –0.19 for this difference. That's okay, but you can avoid negative differences in the sample proportions by having the group with the larger sample proportion serve as the first group (here, females).

Spotting Misleading Confidence Intervals

When the MOE is small, relatively speaking, you would like to say that these confidence intervals provide accurate and credible estimates of their parameters. This is not always the case, however.

Not all estimates are as accurate and reliable as the sources may want you to think. For example, a Web site survey result based on 20,000 hits may have a small MOE according to the formula, but the MOE means nothing if the survey is only given to people who happened to visit that Web site.

In other words, the sample isn't even close to being a random sample (where every sample of equal size selected from the population has an equal chance of being chosen to participate). Nevertheless, such results do get reported, along with their margins of error that make the study seem truly scientific. Beware of these bogus results! (See Chapter 12 for more on the limits of the MOE.)

Before making any decisions based on someone's estimate, do the following:

- Investigate how the statistic was created; it should be the result of a scientific process that results in reliable, unbiased, accurate data.

- Look for a margin of error. If one isn't reported, go to the original source and request it.

- Remember that if the statistic isn't reliable or contains bias, the margin of error will be meaningless.

(See Chapter 16 for evaluating survey data and see Chapter 17 for criteria for good data in experiments.)

Chapter 14

Claims, Tests, and Conclusions

• •

In This Chapter

▶ Testing other people's claims

▶ Using hypothesis tests to weigh evidence and make decisions

▶ Recognizing that your conclusions could be wrong

• •

*Y*ou hear claims involving statistics all the time; the media has no shortage of them:

✔ Twenty-five percent of all women in the United States have varicose veins. (Wow, are some claims better left unsaid, or what?)

✔ Cigarette use in the U.S. continues to drop, with the percentage of all American smokers decreasing by about 2% per year over the last ten years.

✔ A 6-month-old baby sleeps an average of 14 to 15 hours in a 24-hour period. (Yeah, right!)

✔ A name-brand ready-mix pie takes only 5 minutes to make.

In today's age of information (and big money), a great deal rides on being able to back up your claims. Companies that say their products are better than the leading brand had better be able to prove it, or they could face lawsuits. Drugs that are approved by the FDA have to show strong evidence that their products actually work without producing life-threatening side effects. Manufacturers have to make sure their products are being produced according to specifications to avoid recalls, customer complaints, and loss of business.

Although many claims are backed up by solid scientific (and statistically sound) research, others are not. In this chapter, you find out how to use statistics to investigate whether a claim is actually valid and get the lowdown on the process that researchers *should* be using to validate claims that they make.

A *hypothesis test* is a statistical procedure that's designed to test a claim. Before diving into details, I want to give you the big picture of a hypothesis test by showing the main steps involved. These steps are discussed in the following sections:

1. **Set up the null and alternative hypotheses.**

2. Collect good data using a well-designed study (see Chapters 16 and 17).

3. Calculate the test statistic based on your data.

4. Find the *p*-value for your test statistic.

5. Decide whether or not to reject H_o based on your *p*-value.

6. Understand that your conclusion may be wrong, just by chance.

Setting Up the Hypotheses

Typically in a hypothesis test, the claim being made is about a population *parameter* (one number that characterizes the entire population). Because parameters tend to be unknown quantities, everyone wants to make claims about what their values may be. For example, the claim that 25% (or 0.25) of all women have varicose veins is a claim about the proportion (that's the *parameter*) of all women (that's the *population*) who have varicose veins (that's the *variable* — having or not having varicose veins).

Researchers often challenge claims about population parameters. You may hypothesize, for example, that the actual proportion of women who have varicose veins is lower than 0.25, based on your observations. Or you may hypothesize that due to the popularity of high heeled shoes, the proportion may be higher than 0.25. Or if you're simply questioning whether the actual proportion is 0.25, your alternative hypothesis is: "No, it isn't 0.25."

Defining the null

Every hypothesis test contains a set of two opposing statements, or hypotheses, about a population parameter. The first hypothesis is called the *null hypothesis,* denoted H_o. The null hypothesis always states that the population parameter is *equal* to the claimed value. For example, if the claim is that the average time to make a name-brand ready-mix pie is five minutes, the statistical shorthand notation for the null hypothesis in this case would be as follows: $H_o: \mu = 5$. (That is, the population mean is 5 minutes.)

All null hypotheses include an equal sign in them; there are no ≤ or ≥ signs in H_o. Not to cop out or anything, but the reason it's always equal is beyond the scope of this book; let's just say you wouldn't pay me to explain it to you.

What's the alternative?

Before actually conducting a hypothesis test, you have to put two possible hypotheses on the table — the null hypothesis is one of them. But, if the null hypothesis is rejected (that is, there was sufficient evidence against it), what's your alternative going to be? Actually, three possibilities exist for the second (or alternative) hypothesis, denoted H_a. Here they are, along with their shorthand notations in the context of the pie example:

✔ The population parameter is *not equal* to the claimed value (H_a: $\mu \neq 5$).

✔ The population parameter is *greater than* the claimed value (H_a: $\mu > 5$).

✔ The population parameter is *less than* the claimed value (H_a: $\mu < 5$).

Which alternative hypothesis you choose in setting up your hypothesis test depends on what you're interested in concluding, should you have enough evidence to refute the null hypothesis (the claim).

For example, if you want to test whether a company is correct in claiming its pie takes five minutes to make and it doesn't matter whether the actual average time is more or less than that, you use the not-equal-to alternative. Your hypotheses for that test would be H_0: $\mu = 5$ versus H_a: $\mu \neq 5$.

If you only want to see whether the time turns out to be greater than what the company claims (that is, whether the company is falsely advertising its quick prep time), you use the greater-than alternative, and your two hypotheses are H_0: $\mu = 5$ versus H_a: $\mu > 5$.

Finally, say you work for the company marketing the pie, and you think the pie can be made in less than five minutes (and could be marketed by the company as such). The less-than alternative is the one you want, and your two hypotheses would be H_0: $\mu = 5$ versus H_a: $\mu < 5$.

 How do you know which hypothesis to put in H_0 and which one to put in H_a? Typically, the null hypothesis says that nothing new is happening; the previous result is the same now as it was before, or the groups have the same average (their difference is equal to zero). In general, you assume that people's claims are true until proven otherwise. So the question becomes: Can you prove otherwise? In other words, can you show sufficient evidence to reject H_0?

Gathering Good Evidence (Data)

After you've set up the hypotheses, the next step is to collect your evidence and determine whether your evidence goes against the claim made in H_0. Remember, the claim is made about the population, but you can't test the whole population; the best you can usually do is take a sample. As with any other situation in which statistics are being collected, the quality of the data is extremely critical. (See Chapter 3 for ways to spot statistics that have gone wrong.)

Collecting good data starts with selecting a good sample. Two important issues to consider when selecting your sample are avoiding bias and being accurate. To avoid bias when selecting a sample, make it a random sample (one that's got the same chance of being selected as every other possible sample of the same size) and choose a large enough sample size so that the results will be accurate. (See Chapter 11 for more information on accuracy.)

Data is collected in many different ways, but the methods used basically boil down to two: surveys (observational studies) and experiments (controlled studies). Chapter 16 gives all the information you need to design and critique surveys, as well as information on selecting samples properly. In Chapter 17, you examine experiments: what they can do beyond an observational study, the criteria for a good experiment, and when you can conclude cause and effect.

Compiling the Evidence: The Test Statistic

After you select your sample, the appropriate number-crunching takes place. Your null hypothesis (H_o) makes a statement about the population parameter — for example, "The proportion of all women who have varicose veins is 0.25" (in other words, H_o: $p = 0.25$); or the average miles per gallon of a U.S.-built light truck is 27 (H_o: $\mu = 27$). The data you collect from the sample measures the variable of interest, and the statistics that you calculate will help you test the claim about the population parameter.

Gathering sample statistics

Say you're testing a claim about the proportion of women with varicose veins. You need to calculate the proportion of women in your sample who have varicose veins, and that number will be your sample statistic. If you're testing a claim about the average miles per gallon of a U.S.-built light truck, your statistic will be the average miles per gallon of the light trucks in your sample. And knowing you want to measure the variability in average miles per gallon for various trucks, you'll want to calculate the sample standard deviation. (See Chapter 5 for all the information you need on calculating sample statistics.)

Measuring variability using standard errors

After you've calculated all the necessary sample statistics, you may think you're done with the analysis part and ready to make your conclusions — but you're not. The problem is you have no way to put your results into any kind of perspective just by looking at them in their regular units. That's because you know that your results are based only on a sample and that sample results are going to vary. That variation needs to be taken into account, or your conclusions could be completely wrong. (How much do sample results vary? Sample variation is measured by the standard error; see Chapter 11 for more on this.)

Suppose the claim is that the percentage of all women with varicose veins is 25%, and your sample of 100 women had 20% with varicose veins. The standard error for your sample percentage is 4% (according to formulas in Chapter 11), which means that your results are expected to vary by about twice that, or about 8%, according to the Empirical Rule (see Chapter 12). So a difference of 5%, for example, between the claim and your sample result (25% – 20% = 5%) isn't that much, in these terms, because it represents a distance of less than 2 standard errors away from the claim.

However, suppose your sample percentage was based on a sample of 1,000 women, not 100. This decreases the amount by which you expect your results to vary, because you have more information. Again using formulas from Chapter 11, I calculate the standard error to be 0.013 or 1.3%. The margin of error (MOE) is about twice that, or 2.6% on either side. Now a difference of 5% between your sample result (20%) and the claim in H_o (25%) is a more meaningful difference; it's way more than 2 standard errors.

Exactly how meaningful are your results? In the next section, you get more specific about measuring exactly how far apart your sample results are from the claim in terms of the number of standard errors. This leads you to a specific conclusion as to how much evidence you have against the claim in H_o.

Understanding standard scores

The number of standard errors that a statistic lies above or below the mean is called a *standard score* (for example, a *z*-value is a type of standard score; see Chapter 9). In order to interpret your statistic, you need to convert it from original units to a standard score. When finding a standard score, you take your statistic, subtract the mean, and divide the result by the standard error.

In the case of hypothesis tests, you use the value in H_o as the mean. (That's what you go with unless/until you have enough evidence against it.) The standardized version of your statistic is called a *test statistic,* and it's the main component of a hypothesis test. (Chapter 15 contains the formulas for the most common hypothesis tests.)

Calculating and interpreting the test statistic

The general procedure for converting a statistic to a test statistic (standard score) is as follows:

1. **Take your statistic minus the claimed value (the number stated in H_o).**

2. **Divide by the standard error of the statistic.** (Different formulas for standard error exist for different problems; see Chapter 13 for detailed formulas for standard error and Chapter 15 for formulas for various test statistics.)

Your test statistic represents the distance between your actual sample results and the claimed population value, in terms of number of standard errors. In the case of a single population mean or proportion, you know that these standardized distances should at least have an approximate standard normal distribution if your sample size is large enough (see Chapter 11). So, to interpret your test statistic in these cases, you can see where it stands on the standard normal distribution (Z-distribution).

Using the numbers from the varicose veins example in the previous section, the test statistic is found by taking the proportion in the sample with varicose veins, 0.20, subtracting the claimed proportion of all women with varicose veins, 0.25, and then dividing the result by the standard error, 0.04. These calculations give you a test statistic (standard score) of $-0.05 \div 0.04 = -1.25$. This tells you that your sample results and the population claim in H_o are 1.25 standard errors apart; in particular, your sample results are 1.25 standard errors below the claim. Now is this enough evidence to reject the claim? The next section addresses that issue.

Weighing the Evidence and Making Decisions: p-Values

After you find your test statistic, you use it to make a decision about whether to reject H_o. You make this decision by coming up with a number that measures the strength of this evidence (your test statistic) against the claim in H_o. That is, how likely is it that your test statistic could have occurred while the claim was still true? This number you calculate is called the *p-value;* it's the chance that someone could have gotten results as extreme as yours while H_o was still true. Similarly in a jury trial, the jury discusses how likely it is that all the evidence came out the way it did assuming the defendant was innocent.

This section shows all the ins and outs of *p*-values, including how to calculate them and use them to make decisions regarding H_o.

Connecting test statistics and p-values

To test whether a claim in H_o should be rejected (after all, it's all about H_o) you look at your test statistic taken from your sample and see whether you have enough evidence to reject the claim. If the test statistic is large (in either the positive or negative directions), your data is far from the claim; the larger the test statistic, the more evidence you have against the claim. You determine "how far is far" by looking at where your test statistic ends up on the distribution that it came from. When testing one population mean, under certain conditions the distribution of comparison is the standard normal (Z-) distribution, which has a mean of 0 and a standard deviation of 1; I use it throughout this section as an example. (See Chapter 9 to find out more about the Z-distribution.)

If your test statistic is close to 0, or at least within that range where most of the results should fall, then you don't have much evidence against the claim (H_o) based on your data. If your test statistic is out in the tails of the standard normal distribution (see Chapter 9 for more on tails), then your evidence against the claim (H_o) is great; this result has a very small chance of happening if the claim is true. In other words, you have sufficient evidence against the claim (H_o), and you reject H_o.

But how far is "too far" from 0? As long as you have a normal distribution or a large enough sample size, you know that your test statistic falls somewhere on a standard normal distribution (see Chapter 11). If the null hypothesis (H_o) is true, most (about 95%) of the samples will result in test statistics that lie roughly within 2 standard errors of the claim. If H_a is the not-equal-to alternative, any test statistic outside this range will result in H_o being rejected. See Figure 14-1 for a picture showing the locations of your test statistic and their corresponding conclusions. In the next section, you see how to quantify the amount of evidence you have against H_o.

Figure 14-1:
Decisions
for H_a: not-
equal-to.

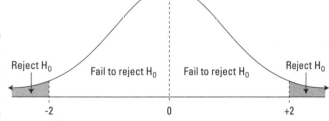

Note that if the alternative hypothesis is the less-than alternative, you reject H_o only if the test statistic falls in the left tail of the distribution (below −1.64). Similarly, if H_a is the greater-than alternative, you reject H_o only if the test statistic falls in the right tail (above 1.64).

Defining a p-value

A *p*-value is a probability associated with your test statistic. It measures the chance of getting results at least as strong as yours if the claim (H_o) were true. In the case of testing the population mean, the farther out your test statistic is on the tails of the standard normal (Z-) distribution, the smaller your *p*-value will be, the less likely your results were to have occurred, and the more evidence you have against the claim (H_o).

Calculating a p-value

To find the *p*-value for your test statistic:

1. **Look up your test statistic on the appropriate distribution — in this case, on the standard normal (Z-) distribution (see the Z-table in the appendix).**

2. **Find the chance that Z is beyond (more extreme than) your test statistic:**

   - If H_a contains a less-than alternative, find the probability that Z is less than your test statistic (that is, look up your test statistic on the Z-table and find its corresponding probability). This is the *p*-value.

   - If H_a contains a greater-than alternative, find the probability that Z is greater than your test statistic (look up your test statistic on the Z-table, find its corresponding probability, and subtract it from one). The result is your *p*-value.

   - If H_a contains a non-equal-to alternative, find the probability that Z is beyond your test statistic and double it. There are two cases:

     If your test statistic is negative, first find the probability that Z is less than your test statistic (look up your test statistic on the Z-table and find its corresponding probability). Then double this probability to get the *p*-value.

     If your test statistic is positive, first find the probability that Z is greater than your test statistic (look up your test statistic on the Z-table, find its corresponding probability, and subtract it from one). Then double this result to get the *p*-value.

Why do you double the probabilities if your H_a contains a non-equal-to alternative? Think of the not-equal-to alternative as the combination of the greater-than alternative and the less-than alternative. If you've got a positive test statistic, its *p*-value only accounts for the greater-than portion of the not-equal-to alternative; double it to account for the less-than portion. (The doubling of one *p*-value is possible because the Z-distribution is symmetric.)

Similarly, if you've got a negative test statistic, its *p*-value only accounts for the less-than portion of the not-equal-to alternative; double it to also account for the greater-than portion.

When testing H_o: $p = 0.25$ versus H_a: $p < 0.25$ in the varicose veins example from the previous section, the *p*-value turns out to be 0.1056. This is because the test statistic (calculated in the previous section) was –1.25, and when you look this number up on the Z-table (in the appendix) you find a probability of 0.1056 of being less than this value. If you had been testing the two-sided alternative, H_a: $p \neq 0.25$, the *p*-value would be 2 * 0.1056, or 0.2112.

If the results are likely to have occurred under the claim, then you fail to reject H_o (like a jury decides not guilty). If the results are unlikely to have occurred under the claim, then you reject H_o (like a jury decides guilty). The cutoff point between rejecting H_o and failing to reject H_o is another whole can of worms that I dissect in the next section (no pun intended).

Making Conclusions

To draw conclusions about H_o (reject or fail to reject) based on a *p*-value, you need to set a predetermined cutoff point where only those *p*-values less than or equal to the cutoff will result in rejecting H_o. This cutoff point is called the *alpha level (α),* or *significance level* for the test. While 0.05 is a very popular cutoff value for rejecting H_o, cutoff points and resulting decisions can vary — some people use stricter cutoffs, such as 0.01, requiring more evidence before rejecting H_o, and others may have less strict cutoffs, such as 0.10, requiring less evidence.

If H_o is rejected (that is, the *p*-value is less than or equal to the predetermined significance level), the researcher can say she's found a statistically significant result. A result is *statistically significant* if it's too rare to have occurred by chance assuming H_o is true. If you get a statistically significant result, you have enough evidence to reject the claim, H_o, and conclude that something different or new is in effect (that is, H_a).

The significance level can be thought of as the highest possible *p*-value that would reject H_o and declare the results statistically significant. Following are the general rules for making a decision about H_o based on a *p*-value:

- ✔ If the *p*-value is less than or equal to your significance level, then it meets your requirements for having enough evidence against H_o; you reject H_o.

- ✔ If the *p*-value is greater than your significance level, your data failed to show evidence beyond a reasonable doubt; you fail to reject H_o.

However, if you plan to make decisions about H_o by comparing the *p*-value to your significance level, you must decide on your significance level ahead of time. It wouldn't be fair to change your cutoff point after you've got a sneak peak at what's happening in the data.

You may be wondering whether it's okay to say "Accept H_o" instead of "Fail to reject H_o." The answer is a big no. In a hypothesis test, you are *not* trying to show whether or not H_o is true (which *accept* implies) — indeed, if you knew whether H_o was true, you wouldn't be doing the hypothesis test in the first place. You're trying to show whether you have enough evidence to say H_o is false, based on your data. Either you have enough evidence to say it's false (in which case you reject H_o) or you don't have enough evidence to say it's false (in which case you fail to reject H_o).

Setting boundaries for rejecting H_o

These guidelines help you make a decision (reject or fail to reject H_o) based on a *p*-value when your significance level is 0.05:

- ✓ If the *p*-value is less than 0.01 (very small), the results are considered highly statistically significant — reject H_o.

- ✓ If the *p*-value is between 0.05 and 0.01 (but not super-close to 0.05), the results are considered statistically significant — reject H_o.

- ✓ If the *p*-value is really close to 0.05 (like 0.051 or 0.049), the results should be considered marginally significant — the decision could go either way.

- ✓ If the *p*-value is greater than (but not super-close to) 0.05, the results are considered non-significant — you fail to reject H_o.

When you hear a researcher say her results are found to be statistically significant, look for the *p*-value and make your own decision; the researcher's pre-determined significance level may be different from yours. If the *p*-value isn't stated, ask for it.

Testing varicose veins

In the varicose veins example in the last section, the *p*-value was found to be 0.1056. This *p*-value is fairly large and indicates very weak evidence against H_o by almost anyone's standards because it's greater than 0.05 and even slightly greater than 0.10 (considered to be a very large significance level). In this case you fail to reject H_o. You didn't have enough evidence to say the proportion of women with varicose veins is less than 0.25 (your alternative hypothesis). This isn't declared to be a statistically significant result.

But say your *p*-value had been something like 0.026. A reader with a personal cutoff point of 0.05 would reject H_o in this case because the *p*-value (of 0.026) is less than 0.05. His conclusion would be that the proportion of women with varicose veins isn't equal to 0.25; according to H_a in this case, you conclude it's less than 0.25, and the results are statistically significant. However, a reader whose significance level is 0.01 wouldn't have enough evidence (based on your sample) to reject H_o because the *p*-value of 0.026 is greater than 0.01. These results wouldn't be statistically significant.

Finally, if the *p*-value turned out to be 0.049 and your significance level is 0.05, you can go by the book and say because it's less than 0.05 you reject H_o, but you really should say your results are marginal, and let the reader decide. (Maybe they can flip a coin or something — "Heads we reject H_o, tails, we don't!")

Assessing the Chance of a Wrong Decision

After you make a decision to either reject H_o or fail to reject H_o, the next step is living with the consequences, in terms of how people respond to your decision.

- ✔ If you conclude that a claim isn't true but it actually *is*, will that result in a lawsuit, a fine, unnecessary changes in the product, or consumer boycotts that shouldn't have happened? It's possible.

- ✔ If you can't disprove a claim that's wrong, what happens then? Will products continue to be made in the same way as they are now? Will no new law be made, no new action taken, because you showed that nothing was wrong? Missed opportunities to blow the whistle have been known to occur.

Whatever decision you make with a hypothesis test, you know there is a chance of being wrong; that's life in the statistics world. Knowing the kinds of errors that can happen and finding out how to curb the chance of them occurring are key.

Making a false alarm: Type-1 errors

Suppose a company claims that its average package delivery time is 2 days, and a consumer group tests this hypothesis, gets a *p*-value of 0.04, and concludes that the claim is false: They believe that the average delivery time is actually more than 2 days. This is a big deal. If the group can stand by its statistics, it has done well to inform the public about the false advertising issue. But what if the group is wrong?

Even if the group bases their study on a good design, collects good data, and makes the right analysis, it can still be wrong. Why? Because its conclusions were based on a sample of packages, not on the entire population. And as Chapter 11 tells you, sample results vary from sample to sample.

Just because the results from a sample are unusual doesn't mean they're impossible. A *p*-value of 0.04 means that the chance of getting your particular test statistic, even if the claim is true, is 4% (less than 5%). You reject H_o in this case because that chance is small. But even a small chance is still a chance!

Perhaps your sample, though collected randomly, just happens to be one of those atypical samples whose result ended up far from what was expected. So, H_o could be true, but your results lead you to a different conclusion. How often does that happen? Five percent of the time (or whatever your given cutoff probability is for rejecting H_o).

Rejecting H_o when you shouldn't is called a *type-1 error*. I don't really like this name, because it seems so nondescript. I prefer to call a type-1 error a *false alarm*. In the case of the packages, if the consumer group made a type-1 error when it rejected the company's claim, they created a false alarm. What's the result? A very angry delivery company, I guarantee that!

To reduce the chance of false alarms, set a low cutoff probability (significance level) for rejecting H_o. Setting it to 5% or 1% will keep the chance of a type-1 error in check.

Missing out on a detection: Type-2 errors

On the other hand, suppose the company really wasn't delivering on its claim. Who's to say that the consumer group's sample will detect it? If the actual delivery time is 2.1 days instead of 2 days, the difference would be pretty hard to detect. If the actual delivery time is 3 days, even a fairly small sample would probably show that something's up. The issue lies with those in-between values, like 2.5 days.

If H_o is indeed false, you want to find out about it and reject H_o. Not rejecting H_o when you should have is called a *type-2 error*. I like to call it a *missed detection*.

Sample size is the key to being able to detect situations where H_o is false and, thus, avoiding type-2 errors. The more information you have, the less variable your results will be (see Chapter 11) and the more ability you have to zoom in on detecting problems that exist with a claim made by H_o.

This ability to detect when H_o is truly false is called the *power* of a test. Power is a pretty complicated issue, but what's important for you to know is that the higher the sample size, the more powerful a test is. A powerful test has a small chance for a type-2 error.

As a preventative measure to minimize the chances of a type-2 error, statisticians recommend that you select a large sample size to ensure that any differences or departures that really exist won't be missed.

Chapter 15

Commonly Used Hypothesis Tests: Formulas and Examples

*F*rom product advertisements to media blitzes on recent medical break-throughs, you often run across claims made about one or more populations. For example, "We promise to deliver our packages in two days or less" or "Two recent studies show that a high-fiber diet may reduce your risk of colon cancer by 20%." Whenever someone makes a claim (also called a *null hypothesis*) about a population (such as all packages, or all adults) you can test the claim by doing what statisticians call a *hypothesis test*.

A hypothesis test involves setting up your *hypotheses* (a claim and its alternative), selecting a sample (or samples), collecting data, calculating the relevant statistics, and using those statistics to decide whether the claim is true.

In this chapter, I outline the formulas used for some of the most common hypothesis tests, explain the necessary calculations, and walk you through some examples.

If you need more background information on hypothesis testing (such as setting up hypotheses, understanding test statistics, *p*-values, significance levels, and type-1 and type-2 errors), just flip to Chapter 14. All the general concepts of hypothesis testing are developed there. This chapter focuses on their application.

Testing One Population Mean

When the variable is numerical (for example, age, income, time, and so on) and only one population or group (such as all U.S. households or all college students) is being studied, you use the hypothesis test in this section to examine or challenge a claim about the population mean. For example, a child psychologist says that the average time that working mothers spend talking to their children is 11 minutes per day, on average. (For dads, the claim is 8 minutes.) The variable — time — is numerical, and the population is all working mothers. Using statistical notation, μ represents the average number of minutes per day that all working mothers spend talking to their children, on average.

The null hypothesis is that the population mean, μ, is equal to a certain claimed value, μ_o. The notation for the null hypothesis is $H_o: \mu = \mu_o$. So the null hypothesis in our example is $H_o: \mu = 11$ minutes, and μ_o is 11. The three possibilities for the alternative hypothesis, H_a, are $\mu \neq 11$, $\mu < 11$, or $\mu > 11$, depending on what you are trying to show. (See Chapter 14 for more on alternative hypotheses.) If you suspect that the average time working mothers spend talking with their kids is more than 11 minutes, your alternative hypothesis would be $H_a: \mu > 11$.

To test the claim, you compare the mean you got from your sample ($\bar{x}$) with the mean shown in H_o (μ_o). To make a proper comparison, you look at the difference between them, and divide by the standard error to take into account the fact that your sample results will vary. (See Chapter 12 for all the info you need on standard error.) This result is your *test statistic*. In the case of a hypothesis test for the population mean, the test statistic turns out (under certain conditions) to be a z-value (a value from the Z-distribution; see Chapter 9).

Then you can look up your test statistic on the appropriate table (in this case, you look it up on the Z-table in the appendix), and find the chance that this difference between your sample mean and the claimed population mean really could have occurred if the claim were true.

The test statistic for testing one population mean (under certain conditions) is

$$z = \frac{\bar{x} - \mu_o}{\sigma/\sqrt{n}}$$

where $\bar{x}$ is the sample mean, σ is the population standard deviation (assume for this case that this number is known), and z is a value on the Z-distribution. To calculate the test statistic, do the following:

1. **Calculate the sample mean, $\bar{x}$.**

2. **Find $\bar{x} - \mu_o$.**

3. **Calculate the standard error: $\sigma/\sqrt{n}$.**

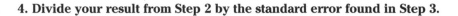

4. Divide your result from Step 2 by the standard error found in Step 3.

The conditions for using this test statistic are that the population standard deviation, σ, is known, and either the population has a normal distribution or the sample size is large enough to use the CLT ($n > 30$); see Chapter 11.

For our example, suppose a random sample of 100 working mothers spend an average of 11.5 minutes per day talking with their children. (Assume prior research suggests the population standard deviation is 2.3 minutes.)

1. We are given that $\bar{x}$ is 11.5, $n = 100$, and σ is 2.3.

2. Take $11.5 - 11 = +0.5$.

3. Take 2.3 divided by the square root of 100 (which is 10) to get 0.23 for the standard error.

4. Divide $+0.5$ by 0.23 to get 2.17. That's your test statistic, which means your sample mean is 2.17 standard errors above the claimed population mean.

The big idea of a hypothesis test is to challenge the claim that's being made about the population (in this case, the population mean); that claim is shown in the null hypothesis, H_o. If you have enough evidence from your sample against the claim, H_o is rejected.

To decide whether you have enough evidence to reject H_o, calculate the p-value by looking up your test statistic (in this case 2.17) on the standard normal (Z-) distribution — see the Z-table in the appendix — and take 1 minus the probability shown. (You subtract from 1 because your H_a is a greater-than hypothesis and the table shows less-than probabilities.)

For this example you look up the test statistic (2.17) on the Z-table and find the (less-than) probability is 0.9850, so the p-value is $1 - 0.9850 = 0.015$. It's quite a bit less than your (typical) significance level 0.05, which means your sample results would be considered unusual if the claim (of 11 minutes) was true. So reject the claim (H_o: $\mu = 11$ minutes). Your results support the alternative hypothesis H_a: $\mu > 11$. According to your data, the child psychologist's claim of 11 minutes per day is too low; the actual average is greater than that.

For information on how to calculate p-values for the less-than or not-equal-to alternatives, also see Chapter 14.

Handling Small Samples and Unknown Standard Deviations: The t-Test

In two cases, you can't use the Z-distribution for a test statistic for one population mean. The first case is where the sample size is small (and by small,

I mean dropping below 30 or so) the second case is when the population standard deviation, σ, is not known, and you have to estimate it using the sample standard deviation, s. In both cases, you have less reliable information on which to base your conclusions, so you have to pay a penalty for this by using a distribution with more variability in the tails than a Z-distribution has. Enter the t-distribution. (See Chapter 10 for all things t-distribution, including its relationship with the Z.)

A hypothesis test for a population mean that involves the t-distribution is called a t-test. The formula for the test statistic in this case is:

$$t_{n-1} = \frac{\bar{x} - \mu_o}{s / \sqrt{n}}, \text{ where } t_{n-1} \text{ is a value from the } t\text{-distribution with } n-1 \text{ degrees}$$

of freedom.

Note it is just like the test statistic for the large sample and/or normal distribution case (see the section "Testing One Population Mean"), except σ is not known, so you substitute the sample standard deviation, s, instead, and use a t-value rather than a z-value.

Because the t-distribution has fatter tails than the Z-distribution, you get a larger p-value from the t-distribution than one that the standard normal (Z-) distribution would have given you for the same test statistic. A bigger p-value means less chance of rejecting H_o. Having less data and/or not knowing the population standard deviation should create a higher burden of proof.

Putting the t-test to work

Suppose a delivery company claims they deliver their packages in 2 days on average, and you suspect it's longer than that. The hypotheses are $H_o: \mu = 2$ versus $H_a: \mu > 2$. To test this claim, you take a random sample of 10 packages and record their delivery times. You find the sample mean is $\bar{x} = 2.3$ days, and the sample standard deviation is 0.35 days. (Because the population standard deviation, σ, is unknown, you estimate it with s, the sample standard deviation.) This is a job for the t-test.

Because the sample size is small ($n = 10$ is much less than 30) and the population standard deviation is not known, your test statistic has a t-distribution. Its degrees of freedom is $10 - 1 = 9$. The formula for the test statistic (referred to as the t-value) is:

$$t_{10-1} = \frac{2.3 - 2.0}{0.35 / \sqrt{10}} = 2.71$$

To calculate the p-value, you look in the row in the t-table (in the appendix) for $df = 9$. Your test statistic (2.71) falls between two values in the row for $df = 9$ in the t-table: 2.26 and 2.82 (rounding to two decimal places). To calculate the *p*-value for your test statistic, find which columns correspond to these two numbers. The number 2.26 appears in the 0.025 column and the number 2.82 appears in the 0.010 column; you now know the *p*-value for your test statistic lies between 0.025 and 0.010 (that is, $0.010 < p\text{-value} < 0.025$).

Using the *t*-table you don't know the exact number for the *p*-value, but because 0.010 and 0.025 are both less than your significance level of 0.05, you reject H_o; you have enough evidence in your sample to say the packages are not being delivered in 2 days, and in fact the average delivery time is more than 2 days.

The *t*-table (in the appendix) doesn't include every possible *t*-value; just find the two values closest to yours on either side, look at the columns they're in, and report your *p*-value in relation to theirs. (If your test statistic is greater than all the *t*-values in the corresponding row of the *t*-table, just use the last one; your *p*-value will be less than its probability.)

Of course you can use statistical software, if available, to calculate exact *p*-values for any test statistic; using software you get 0.012 for the exact *p*-value.

Relating t to Z

The next-to-the-last line of the *t*-table shows the corresponding values from the standard normal (*Z*-) distribution for the probabilities listed on the top of each column. Now choose a column in the table and move down the column looking at the *t*-values. As the degrees of freedom of the *t*-distribution increase, the *t*-values get closer and closer to that row of the table where the *z*-values are.

This confirms a result found in Chapter 10: As the sample size (hence degrees of freedom) increases, the *t*-distribution becomes more and more like the *Z*-distribution, so the *p*-values from their hypothesis tests are virtually equal for large sample sizes. And those sample sizes don't even have to be that large to see this relationship; for $df = 30$ the *t*-values are already very similar to the *z*-values shown in the bottom of the table. These results make sense; the more data you have, the less of a penalty you have to pay. (And of course, you can use computer technology to calculate more exact *p*-values for any *t*-value you like.)

Handling negative t-values

For a less-than alternative hypothesis (H_a: xx < xx), your test statistic would be a negative number (to the left of 0 on the *t*-distribution). In this case, you want to find the percentage below, or to the left of, your test statistic to get your *p*-value. Yet negative test statistics don't appear on the *t*-table (in the appendix).

Not to worry! The percentage to the left (below) a negative *t*-value is the same as the percentage to the right (above) the positive *t*-value, due to symmetry. So to find the *p*-value for your negative test statistic, look up the positive version of your test statistic on the *t*-table, find the corresponding right tail (greater-than) probability, and use that.

For example, suppose your test statistic is –2.7105 with 9 degrees of freedom and H_a is the less-than alternative. To find your *p*-value, first look up +2.7105 on the *t*-table; by the work in the previous section, you know its *p*-value falls between the column headings 0.025 and 0.010. Because the *t*-distribution is symmetric, the *p*-value for –2.7105 also falls somewhere between 0.025 and 0.010. Again you reject H_o because these values are both less than or equal to 0.05.

Examining the not-equal-to alternative

To find the *p*-value when your alternative hypothesis (H_a) is not-equal-to, simply double the probability that you get from the *t*-table when you look up your test statistic. Why double it? Because the *t*-table shows only greater-than probabilities, which are only half the story. To find the *p*-value when you have a not-equal-to alternative, you must add the *p*-values from the less-than and greater-than alternatives. Because the *t*-distribution is symmetric, the less-than and greater-than probabilities are the same, so just double the one you looked up on the *t*-table and you'll have the *p*-value for the not-equal-to alternative.

For example, if your test statistic is 2.7171 and H_a is a not-equal-to alternative, look up 2.7171 on the *t*-table (*df* = 9 again), and you find the *p*-value lies between 0.025 and 0.010, as shown previously. These are the *p*-values for the greater-than alternative. Now double these values to include the less-than alternative and you find the *p*-value for your test statistic lies somewhere between 0.025 ∗ 2 = 0.05 and 0.010 ∗ 2 = 0.020.

Testing One Population Proportion

When the variable is categorical (for example, gender or support/oppose) and only one population or group is being studied (for example, all registered voters), you use the hypothesis test in this section to test a claim about the population proportion. The test looks at the proportion (*p*) of individuals in the population who have a certain characteristic — for example, the proportion of people who carry cellphones. The null hypothesis is H_o: $p = p_o$, where p_o is a certain claimed value of the population proportion, *p*. For example, if the claim is that 70% of people carry cellphones, p_o is 0.70. The alternative hypothesis is one of the following: $p > p_o$, $p < p_o$, or $p \neq p_o$. (See Chapter 14 for more on alternative hypotheses.)

The formula for the test statistic for a single proportion (under certain conditions) is:

$$z = \frac{\hat{p} - p_o}{\sqrt{\dfrac{p_o(1-p_o)}{n}}}$$

where $\hat{p}$ is the proportion of individuals in the sample who have that characteristic and z is a value on the Z-distribution (see Chapter 9). To calculate the test statistic, do the following:

1. **Calculate the sample proportion, $\hat{p}$, by taking the number of people in the sample who have the characteristic of interest (for example, the number of people in the sample carrying cellphones) and dividing that by n, the sample size.**

2. **Find $\hat{p} - p_o$, where p_o is the value in H$_o$.**

3. **Calculate the standard error, $\sqrt{\dfrac{p_o(1-p_o)}{n}}$.**

4. **Divide your result from Step 2 by your result from Step 3.**

To interpret the test statistic, look up your test statistic on the standard normal (Z-) distribution (in the appendix) and calculate the p-value (see Chapter 14 for more on p-value calculations).

The conditions for using this test statistic are that $np_o \geq 10$ and $n(1-p_o) \geq 10$ (see Chapter 9 for details).

For example, suppose Cavifree claims that four out of five dentists recommend Cavifree toothpaste to their patients. In this case, the population is all dentists, and p is the proportion of all dentists who recommended Cavifree. The claim is that p is equal to "four out of five," or p_o is $4 \div 5 = 0.80$. You suspect that the proportion is actually less than 0.80. Your hypotheses are H$_o$: $p = 0.80$ versus H$_a$: $p < 0.80$.

Suppose that 151 out of your sample of 200 dental patients reported receiving a recommendation for Cavifree from their dentist. To find the test statistic for these results, follow these steps:

1. Start with $\hat{p} = \dfrac{151}{200} = 0.755$ and $n = 200$.

2. Because $p_o = 0.80$, take $0.755 - 0.80 = -0.045$ (the numerator of the test statistic).

3. Next, the standard error equals $\sqrt{\dfrac{0.80(1-0.80)}{200}} = 0.028$ (the denominator of the test statistic).

4. The test statistic is $\dfrac{-0.045}{0.028} = -1.61$.

Because the resulting test statistic is negative, it means your sample results are −1.61 standard errors below (less than) the claimed value for the population. How often would you expect to get results like this if H_o were true? The chance of being at or beyond (in this case less than) −1.61 is 0.0537. (Keep the negative with the number and look up −1.61 in the Z-table in the appendix.) This result is your p-value because H_a is a less-than hypothesis. (See Chapter 14 for more on this.)

Because the p-value is greater than 0.05 (albeit not by much), you don't have quite enough evidence for rejecting H_o. You conclude that the claim that 80% of dentists recommend Cavifree can't be rejected, according to your data. However, it's important to report the actual p-value too, so others can make their own decisions.

The letter p is used two different ways in this chapter: p-value and p. The letter p by itself indicates the population proportion, not the p-value. Don't get confused. Whenever you report a p-value, be sure you add *-value* so it's not confused with p, the population proportion.

Comparing Two (Independent) Population Averages

When the variable is numerical (for example, income, cholesterol level, or miles per gallon) and two populations or groups are being compared (for example, men versus women), you use the steps in this section to test a claim about the difference in their averages. (For example, is the difference in the population means equal to zero, indicating their means are equal?) Two independent (totally separate) random samples need to be selected, one from each population, in order to collect the data needed for this test.

The null hypothesis is that the two population means are the same; in other words, that their difference is equal to 0. The notation for the null hypothesis is $H_o: \mu_1 = \mu_2$, where μ_1 represents the mean of the first population and μ_2 represents the mean of the second population.

You can also write the null hypothesis as $H_o: \mu_1 - \mu_2 = 0$, emphasizing the idea that their difference is equal to zero if the means are the same.

The formula for the test statistic comparing two means (under certain conditions) is:

$$z = \frac{(\bar{x}_1 - \bar{x}_2) - 0}{\sqrt{\dfrac{\sigma_1^2}{n_1} + \dfrac{\sigma_2^2}{n_2}}}$$

To calculate it, do the following:

1. **Calculate the sample means $\bar{x}_1$ and $\bar{x}_2$. (Assume the population standard deviations, σ_1 and σ_2 are given.) Let n_1 and n_2 represent the two sample sizes (they need not be equal).**

   See Chapter 5 for these calculations.

2. **Find the difference between the two sample means: $\bar{x}_1 - \bar{x}_2$.**

   Because $\mu_1 - \mu_2$ is equal to 0 if H_o is true, it doesn't need to be included in the numerator of the test statistic. However, if the difference they are testing is any value other than 0, you subtract that value in the numerator of the test statistic.

3. **Calculate the standard error using the following equation:**

$$\sqrt{\frac{\sigma_1^2}{n_1} + \frac{\sigma_2^2}{n_2}}$$

4. **Divide your result from Step 2 by your result from Step 3.**

To interpret the test statistic, add the following two steps to the list:

5. **Look up your test statistic on the standard normal (Z-) distribution (see the Z-table in the appendix) and calculate the p-value.**

   (See Chapter 14 for more on p-value calculations.)

6. **Compare the p-value to your significance level, such as 0.05. If it's less than or equal to 0.05, reject H_o. Otherwise, fail to reject H_o.**

   (See Chapter 14 for the details on significance levels.)

The conditions for using this test are that the two population standard deviations are known and either both populations have a normal distribution or both sample sizes are large enough for the Central Limit Theorem (see Chapter 11).

For example, suppose you want to compare the absorbency of two brands of paper towels (call the brands Stats-absorbent and Sponge-o-matic). You can make this comparison by looking at the average number of ounces each brand can absorb before being saturated. H_o says the difference between the average absorbencies is 0 (nonexistent), and H_a says the difference is not 0. In other words, one brand is more absorbent than the other. Using statistical notation, you have $H_o = \mu_1 - \mu_2 = 0$ versus $H_a = \mu_1 - \mu_2 \neq 0$. Here, you have no indication of which paper towel may be more absorbent, so the not-equal-to alternative is the one to use (see Chapter 14).

Suppose you select a random sample of 50 paper towels from each brand and measure the absorbency of each paper towel. Suppose the average absorbency of Stats-absorbent (x_1) for your sample is 3 ounces, and assume the population standard deviation is 0.9 ounces. For Sponge-o-matic (x_2), the average absorbency is 3.5 ounces according to your sample; assume the population standard deviation is 1.2 ounces. Carry out this hypothesis test by following the 6 steps listed above:

1. Given the above information, you know $\bar{x}_1 = 3$, $\sigma_1 = 0.9$, $\bar{x}_2 = 3.5$, $\sigma_2 = 1.2$, $n_1 = 50$, and $n_2 = 50$.

2. The difference between the sample means for (Stats-absorbent – Sponge-o-matic) is $\bar{x}_1 - \bar{x}_2 = (3 - 3.5) = -0.5$ ounces. (A negative difference simply means that the second sample mean was larger than the first.)

3. The standard error is $\sqrt{\dfrac{\sigma_1^2}{n_1} + \dfrac{\sigma_2^2}{n_2}} = \sqrt{\dfrac{0.9^2}{50} + \dfrac{1.2^2}{50}} = \sqrt{\dfrac{0.81}{50} + \dfrac{1.44}{50}} = 0.2121$.

4. Divide the difference, –0.5, by the standard error, 0.2121, which gives you –2.36. This is your test statistic.

5. To find the p-value, look up –2.36 on the standard normal (Z-) distribution — see the Z-table in the appendix. The chance of being beyond, in this case to the left of, –2.36 is equal to 0.0091. Because H_a is a not-equal-to alternative, you double this percentage to get $2 * 0.0091 = 0.0182$, your p-value. (See Chapter 14 for more on the not-equal-to alternative.)

6. This p-value is quite a bit less than 0.05. That means you have fairly strong evidence to reject H_o.

Your conclusion is that a statistically significant difference exists between the absorbency levels of these two brands of paper towels, based on your samples. And Sponge-o-matic comes out on top, because it has a higher average. (Stats-absorbent minus Sponge-o-matic being negative means Sponge-o-matic had the higher value.)

If one or both of your samples happen to be under 30 in size, you use the t-distribution (with degrees of freedom equal to $n_1 - 1$ or $n_2 - 1$, whichever is smaller) to look up the p-value. If the population standard deviations, σ_1 and σ_2, are unknown, you use the sample standard deviations s_1 and s_2 instead, and you use the t-distribution with the abovementioned degrees of freedom. (See Chapter 10 for more on the t-distribution.)

Testing for an Average Difference (The Paired t-Test)

You can test for an average difference using the test in this section when the variable is numerical (for example, income, cholesterol level, or miles per gallon) and the individuals in the sample are either paired up in some way according to relevant variables such as age or perhaps weight, or the same people are used twice (for example, using a pre-test and post-test). Paired tests are typically used for studies in which someone is testing to see whether a new treatment, technique, or method works better than an existing method, without having to worry about other factors about the subjects that may influence the results (see Chapter 17 for details).

The average difference (tested in this section) isn't the same as the difference in the averages (tested in the previous section):

- ✔ With the difference in averages, you compare the difference in the means of two separate samples to test the difference in the means of two different populations.

- ✔ With the average difference, you match up the subjects so they are thought of as coming from a single population, and the set of differences measured for each subject (for example, pre-test versus post-test) are thought of as one sample. The hypothesis test then boils down to a test for one population mean (as I explain earlier in this chapter).

For example, suppose a researcher wants to see whether teaching students to read using a computer game gives better results than teaching with a tried-and-true phonics method. She randomly selects 20 students and puts them into 10 pairs according to their reading readiness level, age, IQ, and so on. She randomly selects one student from each pair to learn to read via the computer game method (abbreviated CM), and the other learns to read using the phonics method (abbreviated PM). At the end of the study, each student takes the same reading test. The data are shown in Table 15-1.

Table 15-1   Reading Scores for Computer Game Method versus Phonics Method

Student Pair	Computer Method	Phonics Method	Difference (CM – PM)
1	85	80	+5
2	80	80	0
3	95	88	+7
4	87	90	−3
5	78	72	+6
6	82	79	+3
7	57	50	+7
8	69	73	−4
9	73	78	−5
10	99	95	+4

The original data are in pairs, but you're really interested only in the difference in reading scores (computer reading score minus phonics reading score) for each pair, not the reading scores themselves. So the *paired differences* (the differences in the pairs of scores) are your new data set. See their values in the last column of Table 15-1.

By examining the differences in the pairs of observations, you really only have a single data set, and you only have a hypothesis test for one population mean. In this case the null hypothesis is that the mean (of the paired differences) is 0, and the alternative hypothesis is that the mean (of the paired differences) is > 0.

If the two reading methods are the same, the average of the paired differences should be 0. If the computer method is better, the average of the paired differences should be positive; the computer reading score is larger than the phonics score.

The notation for the null hypothesis is $H_o: \mu_d = 0$, where μ_d is the mean of the paired differences for the population. (The *d* in the subscript just reminds you that you're working with the paired differences.)

The formula for the test statistic for paired differences is $t_{n-1} = \dfrac{\bar{d} - 0}{s_d / \sqrt{n_d}}$, where

$\bar{d}$ is the average of all the paired differences found in the sample, and t_{n-1} is a value on the *t*-distribution with $n_d - 1$ degrees of freedom (see Chapter 10).

You use a *t*-distribution here because in most matched-pairs experiments the sample size is small and/or the population standard deviation σ_d is unknown, so it's estimated by s_d. (See Chapter 10 for more on the *t*-distribution.)

To calculate the test statistic for paired differences, do the following:

1. **For each pair of data, take the first value in the pair minus the second value in the pair to find the paired difference.**

   Think of the differences as your new data set.

2. **Calculate the mean, $\bar{d}$, and the standard deviation, s_d, of all the differences.**

3. **Letting n_d represent the number of paired differences that you have, calculate the standard error:**

   $$s_d / \sqrt{n_d}$$

4. **Divide $\bar{d}$ by the standard error from Step 3.**

Because μ_d is equal to 0 if H_o is true, it doesn't really need to be included in the formula for the test statistic. As a result, you sometimes see the test statistic written like this:

$$\frac{\bar{d} - 0}{s_d / \sqrt{n_d}} = \frac{\bar{d}}{s_d / \sqrt{n_d}}$$

For the reading scores example, you can use the preceding steps to see whether the computer method is better in terms of teaching students to read.

To find the statistic, follow these steps:

1. **Calculate the differences for each pair (they're shown in column 4 of Table 15-1).**

   Notice that the sign on each of the differences is important; it indicates which method performed better for that particular pair.

2. **Calculate the mean and standard deviation of the differences from Step 1.**

   My calculations found the mean of the differences, $\bar{d} = 2$, and the standard deviation is $s_d = 4.64$. Note that $n_d = 10$ here.

3. **The standard error is $\frac{4.64}{\sqrt{10}} = 1.47$.**

   (Remember that here, n_d is the number of pairs, which is 10.)

4. **Take the mean of the differences (Step 2) divided by the standard error of 1.47 (Step 3) to get 1.36, the test statistic.**

Is the result of Step 4 enough to say that the difference in reading scores found in this experiment applies to the whole population in general? Because the population standard deviation, σ, is unknown and you estimated it with the sample standard deviation (s), you need to use the t-distribution rather than the Z-distribution to find your p-value (see the section "Handling Small Samples and Unknown Standard Deviations: The t-Test," earlier in this chapter). Using the t-table (in the appendix) you look up 1.36 on the t-distribution with $10 - 1 = 9$ degrees of freedom to calculate the p-value.

The p-value in this case is greater than 0.05 because 1.36 is smaller than (or to the left of) the value of 1.38 on the table, and therefore its p-value is more than 0.10 (the p-value for the column heading corresponding to 1.38).

Because the p-value is greater than 0.05, you fail to reject H_o; you don't have enough evidence that the mean difference in the scores between the computer method and the phonics method is significantly greater than 0.

However, that doesn't necessarily mean a real difference isn't present in the population of all students. But the researcher can't say the computer game is a better reading method based on this sample of 10 students. (See Chapter 14 for information on the power of a hypothesis test and its relationship to sample size.)

In many paired experiments, the data sets are small due to costs and time associated with doing these kinds of studies. That means the *t*-distribution (see the *t*-table in the appendix) is often used instead of the standard normal (*Z*-) distribution (the *Z*-table in the appendix) when figuring out the *p*-value.

Comparing Two Population Proportions

This test is used when the variable is categorical (for example, smoker/nonsmoker, Democrat/Republican, support/oppose an opinion, and so on) and you're interested in the proportion of individuals with a certain characteristic — for example, the proportion of smokers. In this case, two populations or groups are being compared (such as the proportion of female smokers versus male smokers).

In order to conduct this test, two independent (separate) random samples need to be selected, one from each population. The null hypothesis is that the two population proportions are the same; in other words, that their difference is equal to 0. The notation for the null hypothesis is $H_o: p_1 = p_2$, where p_1 is the proportion from the first population, and p_2 is the proportion from the second population.

Stating in H_o that the two proportions are equal is the same as saying their difference is zero. If you start with the equation $p_1 = p_2$ and subtract p_2 from each side, you get $p_1 - p_2 = 0$. So you can write the null hypothesis either way.

The formula for the test statistic comparing two proportions (under certain conditions) is

$$z = \frac{(\hat{p}_1 - \hat{p}_2) - 0}{\sqrt{\hat{p}(1-\hat{p})\left(\frac{1}{n_1} + \frac{1}{n_2}\right)}}$$

where $\hat{p}_1$ is the proportion in the first sample with the characteristic of interest, $\hat{p}_2$ is the proportion in the second sample with the characteristic of interest, $\hat{p}$ is the proportion in the combined sample (all the individuals in the first and second samples together) with the characteristic of interest, and z is a value on the *Z*-distribution (see Chapter 9). To calculate the test statistic, do the following:

1. Calculate the sample proportions $\hat{p}_1$ and $\hat{p}_2$ for each sample. Let n_1 and n_2 represent the two sample sizes (they don't need to be equal).

2. Find the difference between the two sample proportions, $\hat{p}_1 - \hat{p}_2$.

3. Calculate the overall sample proportion $\hat{p}$, the total number of individuals from both samples who have the characteristic of interest (for example, the total number of smokers, male or female, in the sample), divided by the total number of individuals from both samples ($n_1 + n_2$).

4. Calculate the standard error:

$$\sqrt{\hat{p}(1-\hat{p})\left(\frac{1}{n_1}+\frac{1}{n_2}\right)}$$

5. Divide your result from Step 2 by your result from Step 4. This answer is your test statistic.

To interpret the test statistic, look up your test statistic on the standard normal (Z-) distribution (the Z-table in the appendix) and calculate the p-value, then make decisions as usual (see Chapter 14 for more on p-values).

Consider those drug ads that pharmaceutical companies put in magazines. The front page of an ad shows a serene picture of the sun shining, flowers blooming, people smiling — their lives changed by the drug. The company claims that its drugs can reduce allergy symptoms, help people sleep better, lower blood pressure, or fix whichever other ailment it's targeted to help. The claims may sound too good to be true, but when you turn the page to the back of the ad, you see all the fine print where the drug company justifies how it's able to make its claims. (This is typically where statistics are buried!) Somewhere in the tiny print, you'll likely find a table that shows adverse effects of the drug when compared to a *control group* (subjects who take a fake drug), for fair comparison to those who actually took the real drug (the *treatment group;* see Chapter 17 for more on this).

For example, Adderall, a drug for attention deficit hyperactivity disorder (ADHD), reported that 26 of the 374 subjects (7%) who took the drug experienced vomiting as a side effect, compared to 8 of the 210 subjects (4%) who were on a *placebo* (fake drug). Note that patients didn't know which treatment they were given. In the sample, more people on the drug experienced vomiting, but is this percentage enough to say that the entire population on the drug would experience more vomiting? You can test it to see.

In this example, you have $H_0: p_1 - p_2 = 0$ versus $H_0: p_1 - p_2 > 0$, where p_1 represents the proportion of subjects who vomited using Adderall, and p_2 represents the proportion of subjects who vomited using the placebo.

Why does H_a contain a ">" sign and not a "<" sign? H_a represents the scenario in which those taking Adderall experience more vomiting than those on the placebo — that's something the FDA (and any candidate for the drug) would want to know about. But the order of the groups is important, too. You want to set it up so the Adderall group is first, so that when you take the Adderall proportion minus the placebo proportion, you get a positive number if H_a is true. If you switch the groups, the sign would have been negative.

Now calculate the test statistic:

1. First, determine that

   $$\hat{p}_1 = \frac{26}{374} = 0.070 \text{ and } \hat{p}_2 = \frac{8}{210} = 0.038$$

   The sample sizes are n_1 = 374 and n_2 = 210, respectively.

2. Take the difference between these sample proportions to get
   $\hat{p}_1 - \hat{p}_2 = 0.070 - 0.038 = 0.032$.

3. Calculate the overall sample proportion to get $\hat{p} = \frac{26+8}{374+210} = 0.058$.

4. The standard error is $\sqrt{0.058(1-0.058)\left(\frac{1}{374} + \frac{1}{210}\right)} = 0.020$.

5. Finally, the test statistic is 0.032 ÷ 0.020 = 1.60. Whew!

The *p*-value is the percentage chance of being at or beyond (in this case to the right of) 1.60, which is 1 – 0.9452 = 0.0548. This *p*-value is just slightly greater than 0.05, so, technically, you don't have quite enough evidence to reject H_o. That means that according to your data, vomiting is not experienced any more by those taking this drug when compared to a placebo.

A *p*-value that's very close to that magical but somewhat arbitrary significance level of 0.05 is what statisticians call a *marginal result*. In the preceding example, because the *p*-value of 0.0548 is close to the borderline between accepting and rejecting H_o, it's generally viewed as a marginal result and should be reported as such.

The beauty of reporting a *p*-value is that you can look at it and decide for yourself what you should conclude. The smaller the *p*-value, the more evidence you have against H_o, but how much evidence is enough evidence? Each person is different. If you come across a report from a study in which someone found a statistically significant result, and that result is important to you, ask for the *p*-value so that you can make your own decision. (See Chapter 14 for more.)

Part V
Statistical Studies and the Hunt for a Meaningful Relationship

The 5th Wave   By Rich Tennant

"Okay – let's play the statistical probabilities of this situation. There are 4 of us and 1 of him. Phillip will probably start screaming, Nora will probably faint, you'll probably yell at me for leaving the truck open, and there's a good probability I'll run like a weenie if he comes toward us."

In this part . . .

Many statistics you hear and see each day are based on the results of surveys, experiments, and observational studies. Unfortunately, you can't believe everything you read or hear.

In this part, you look at what actually happens behind the scenes of these studies — how they are designed and conducted and how the data is (supposed to be) collected — so that you'll be able to spot misleading results. You also see what's needed to conduct your own study correctly and effectively.

You also analyze data from good studies to look for relationships between two variables, where both variables are categorical (using two-way tables) or both are numerical (using correlation and regression). In addition, you see how to make proper conclusions and spot problems.

Chapter 16

Polls, Polls, and More Polls

● ●

In This Chapter

▶ Realizing the impact of polls and surveys

▶ Going behind the scenes of polls and surveys

▶ Detecting biased and inaccurate survey results

● ●

Surveys are all the rage amid today's information explosion. Everyone wants to know how the public feels about issues from prescription drug prices and methods of disciplining children to approval ratings of the president and ratings of reality TV shows. Polls and surveys are a big part of American life; they're a vehicle for quickly getting information about how you feel, what you think, and how you live your life, and they're a means of quickly disseminating information about important issues. Surveys highlight controversial topics, raise awareness, make political points, stress the importance of an issue, and educate or persuade the public.

 Survey results can be powerful, because when many people hear that "such and such percentage of the American people do this or that," they accept these results as the truth, and then make decisions and form opinions based on that information. But in fact, many surveys *don't* provide correct, complete, or even fair or balanced information.

In this chapter, I discuss the impact of surveys and how they're used, and I take you behind the scenes of how surveys are designed and conducted so you know what to watch for when examining survey results and how to run your own surveys right. I also talk about how to interpret survey results and how to spot biased and inaccurate information, so that you can determine for yourself which results to believe and which to ignore.

Recognizing the Impact of Polls

A *survey* is an instrument that collects data through questions and answers. It is used to gather information about the opinions, behaviors, demographics, lifestyles, and other reportable characteristics of the population of interest.

What's the difference between a poll and a survey? Statisticians don't make a clear distinction between the two, but I've noticed that what people call a *poll* is typically a short survey containing only a few questions (maybe that's how researchers get more people to respond — they call it a poll rather than a survey!). But for all intents and purposes, surveys and polls are the same thing.

You come into contact with surveys and their results on a daily basis. Compared to other types of studies, such as medical experiments, some surveys can be relatively easy to conduct. They provide quick results that can often make interesting headlines in newspapers or eye-catching stories in magazines. People connect with surveys because they feel that survey results represent the opinions of people just like themselves (even though they may never have been asked to participate in a survey). And many people enjoy seeing how other people feel, what they do, where they go, and what they care about. Looking at survey results makes people feel linked with a bigger group, somehow. That's what *pollsters* (the people who conduct surveys) bank on, and that's why they spend so much time doing surveys and polls and reporting the results of this research.

Getting to the source

Who conducts surveys these days? Pretty much anyone and everyone who has a question to ask. Some of the groups that conduct polls and report the results include the following:

- News organizations
- Political parties and candidates running for office
- Professional polling organizations (such as the Gallup Organization, the Harris Poll, Zogby International, and the National Opinion Research Center [NORC])
- Representatives of magazines, TV shows, and radio programs
- Professional research organizations (like the American Medical Association, Smithsonian Institution, and Pew Research Center for the People and the Press)
- Special-interest groups (such as the National Rifle Association, Greenpeace, and American Civil Liberties Union)
- Academic researchers
- The United States government
- Joe Six-Pack (who can easily conduct his own survey on the Internet)

Ranking the worst cars of the millennium

You may be familiar with a radio show called *Car Talk* that's typically aired Saturday mornings on National Public Radio and is hosted by "Click and Clack," two brothers in Cambridge, Massachusetts, who offer wise and wacky advice to callers with strange car problems. The show's Web site regularly offers "just for fun" surveys on a wide range of car-related topics, such as, "Who has bumper stickers on their cars, and what do they say?" One of their surveys asked the question, "What do you think was the worst car of the millennium?" Thousands upon thousands of folks responded with their votes — but, of course, these folks don't represent all car owners. They represent only those who listen to the radio show, logged on to the Web site, and answered the survey question.

Just so you won't be left hanging (and I know you're dying to find out!), the results of the survey are shown in the following table. Although you may not be old enough to remember some of these vehicles, it is certainly an easy exercise to search the Internet for pictures and stories about them galore. (Remember, though, that these results represent only the opinions of *Car Talk* fans who took the time to get to the Web site and take the survey.) Notice that the percentages won't add up to 100% because the results in the table represent only the top ten vote-getters.

Rank	Type of Car	Percentage of Votes
1	Yugo	33.7%
2	Chevy Vega	15.8%
3	Ford Pinto	12.6%
4	AMC Gremlin	8.5%
5	Chevy Chevette	7.0%
6	Renault LeCar	4.3%
7	Dodge Aspen / Plymouth Volare	4.1%
8	Cadillac Cimarron	4.0%
9	Renault Dauphine	3.6%
10	Volkswagen (VW) Bus	2.7%

Some surveys are just for fun, and others are more serious. Be sure to check the source of any serious survey in which you're asked to participate and for which you're given results. Groups that have a special interest in the results should either hire an independent organization to conduct (or at least to review) the survey, or they should offer copies of the survey questions to the public. Groups should also disclose in detail how the survey was designed and conducted, so that the public can make an informed decision about the credibility of the results.

Surveying what's hot

The topics of many surveys are driven by current events, issues, and areas of interest; after all, timeliness and relevance to the public are two of the most attractive qualities of any survey. Here are just a few examples of some of the subjects being brought to the surface by today's surveys, along with some of the results being reported:

- ✔ Does celebrity activism influence the political opinions of the American public? (Over 90% of the American public says no, according to CBS News.)

- ✔ What percentage of Americans have dated a co-worker? (A whopping 40% have, according to a career networking Web site.)

- ✔ How many patients surf the Web to find health-related information? (55% do, according to a national medical journal.)

When you read the preceding survey results, do you find yourself thinking about what the results mean to you, rather than first asking yourself whether the results are valid? Some of the preceding survey results are more valid and accurate than others, and you should think about whether to believe the results first, before accepting them without question. Nationally known polling and research organizations such as those mentioned in the previous section are credible sources, as well as journals that are *peer-reviewed* (meaning all papers published in the journal have been reviewed by others in the field and passed a certain set of standards). And the U.S. government does a good job with their data collection as well. If you are not familiar with a group conducting a survey and the results are important to you, check out the source.

Impacting lives

Whereas some surveys are just fun to look at and think about, other surveys can have a direct impact on your life or your workplace. These life-decision surveys need to be closely scrutinized before action is taken or important decisions are made. Surveys at this level can cause politicians to change or create new laws, motivate researchers to work on the latest problems, encourage manufacturers to invent new products or change business policies and practices, and influence people's behavior and ways of thinking. The following are some examples of survey results that can impact you:

- ✔ **Children's healthcare suffers:** A survey of 400 pediatricians by the Children's National Medical Center in Washington, D.C., reported that pediatricians spend, on average, only 8 to 12 minutes with each patient.

✔ **Teens drink more:** According to the 2009 Partnership Attitude Tracking Study, conducted by the Partnership for a Drug-Free America, the number of teens in grades 9 through 12 that use alcohol has grown by 4% (from 35% in 2008 to 39% in 2009), reversing the downward trend experienced in the ten years prior to the survey.

Always look at how researchers define the terms they're using to collect their data. In the above example, how did they define "alcohol use"? Does it count if the teenager tried alcohol once? Does it mean they drink alcohol on a consistent basis? Results can be misleading if the range of what or who gets counted is too wide. Find out what questions were actually asked when the data was collected.

✔ **Crimes go unreported:** The U.S. Bureau of Justice Crime Victimization Survey concludes that only 49.4% of violent crimes were reported to police. The reasons victims gave for not reporting crimes to the police are listed in Table 16-1.

Table 16-1   Reasons Victims Didn't Report Violent Crimes

Reason for Not Reporting	*Percentage of Victims*
Considered it to be a personal matter	19.2%
The offender was not successful/didn't complete the crime	15.9%
Reported the crime to another official	14.7%
Didn't consider the crime to be important enough	5.5%
Didn't think police would want to be bothered	5.3%
Lack of proof	5.0%
Fear of reprisal	4.6%
Too inconvenient/time consuming to report it	3.9%
Thought police would be biased/ineffective	2.7%
Property stolen had no ID number	0.5%
Not aware that a crime occurred until later	0.4%
Other reasons	22.3%

The most frequently given reason for not reporting a violent crime to the police was that the victim considered it to be a personal matter (19.2%). Note that almost 12% of the reasons relate to perception of the reporting process itself (for example, that it would take too much time or that the police would be bothered, biased, or ineffective).

By the way, did you notice how large the "Other reasons" category is? This large, unexplained percentage indicates that the survey can be more specific and/or more research can be done regarding why crime victims don't report crimes. Maybe the victims themselves aren't even sure.

Behind the Scenes: The Ins and Outs of Surveys

Surveys and their results are a part of your daily experience, and you use these results to make decisions that affect your life. (Some decisions may even be life changing.) Looking at surveys with a critical eye is important. Before taking action or making decisions based on survey results, you must determine whether those results are credible, reliable, and believable. A good way to begin developing these detective skills is to go behind the scenes and see how surveys are designed, developed, implemented, and analyzed.

The survey process can be broken down into a series of ten steps:

1. **Clarify the purpose of your survey.**
2. **Define the target population.**
3. **Choose the type and timing of the survey.**
4. **Design the introduction with ethics in mind.**
5. **Formulate the questions.**
6. **Select the sample.**
7. **Carry out the survey.**
8. **Follow up, follow up, and follow up.**
9. **Organize and analyze the data.**
10. **Draw conclusions.**

Each step presents its own set of special issues and challenges, but each step is critical in terms of producing survey results that are fair and accurate. This sequence of steps helps you design, plan, and implement a survey, but it can also be used to critique someone else's survey, if those results are important to you.

Planning and designing a survey

The purpose of a survey is to answer questions about a target population. The *target population* is the entire group of individuals that you're interested in drawing conclusions about. In most situations, surveying the entire target population (that is, conducting a full-blown *census*) is impossible because researchers would have to spend too much time or money to do so. Usually, the best you can do is to select a sample of individuals from the target population, survey those individuals, then draw conclusions about the target population based on the data from that sample.

Sounds easy, right? Wrong. Many potential problems arise after you realize that you can't survey everyone in the entire target population. Then, after a sample is selected, many researchers aren't sure what to do to get the data they need. Unfortunately, many surveys are conducted without taking the time needed to think through these issues, resulting in errors, misleading results, and wrong conclusions. In the following sections, I give specifics for the first five steps in the survey process.

Clarifying the purpose of your survey

This sounds like it should just be common sense, but in reality, many surveys have been designed and carried out that never met their purpose, or that met only some of the objectives, but not all of them. Getting lost in the questions and forgetting what you're really trying to find out is easy to do. In stating the purpose of a survey, be as specific as possible. Think about the types of conclusions you would want to make if you were to write a report, and let that help you determine your goals for the survey.

Lots of researchers can't see the forest for the trees. If a restaurant manager wants to determine and compare satisfaction rates for her customers, she needs to think ahead about what kinds of comparisons she wants to make and what information she wants to be able to report on. Questions that pinpoint when the customers came into the restaurant (date and time), or even what table they were at, are relevant. And if she wants to compare satisfaction rates for, say, adults versus families, she needs to ask how many people were in the party and how many were children. But if she simply asks a couple of questions on satisfaction or throws in every question she can think of, without considering in advance why she needs the information, she may end up with more questions than answers.

The more specific you can be about the purpose of the survey, the more easily you can design questions that meet your objectives, and the better off you'll be when you need to write your report.

Defining the target population

Suppose, for example, that you want to conduct a survey to determine the extent to which people send and receive personal e-mail in the workplace. You may think that the target population is e-mail users in the workplace. However, you want to determine the *extent* to which personal e-mail is used in the workplace, so you can't just ask e-mail users, or your results would be biased against those who don't use e-mail in the workplace. But should you also include those who don't even have access to a computer during their workday? (See how fast surveys can get tricky?)

The target population that probably makes the most sense here is all the people who use Internet-connected computers in the workplace. Everyone in this group at least has access to e-mail, though only some of those with access to e-mail in the workplace actually use it, and of those who use it, only some use it for personal e-mail. (And that's what you want to find out — how much they use e-mail for that purpose.)

You need to be clear in your definition of the target population. Your definition is what helps you select the proper sample, and it also guides you in your conclusions, so that you don't overgeneralize your results. If the researcher didn't clearly define the target population, this can be a sign of other problems with the survey.

Choosing the type and timing of the survey

The next step in designing your survey is to choose what type of survey is most appropriate for the situation at hand. Surveys can be done over the phone, through the mail, with door-to-door interviews, or over the Internet. However, not every type of survey is appropriate for every situation. For example, suppose you want to determine some of the factors that relate to illiteracy in the United States. You wouldn't want to send a survey through the mail, because people who can't read won't be able to take the survey. In that case, a telephone interview is more appropriate.

Choose the type of survey that's most appropriate for the target population, in terms of getting the most truthful and informative data possible. You also have to keep in mind the budget you have to work with; door-to-door interviews are more expensive than phone surveys, for example. When examining the results of a survey, be sure to look at whether the type of survey used is most appropriate for the situation, keeping budget considerations in mind.

Next you need to decide when to conduct the survey. In life, timing is everything, and the same goes for surveys. Current events shape people's opinions all the time, and although some pollsters try to determine how people feel about those events, others take advantage of events, especially negative ones, and use them as political platforms or as fodder for headlines and controversy. For example, surveys about gun control often come up after a shooting takes place. Also take note of other events that were going on at the time of the survey; for example, people may not want to answer their phones during the Super Bowl, on election night, during the Olympics, or around holidays. Improper timing can lead to bias.

In addition to the date, the time of day is also important. If you conduct a telephone survey to get people's opinions on stress in the workplace and you call them at home between the hours of 9 a.m. and 5 p.m., you're going to have bias in your results; those are the hours when the majority of people are at work (busy being stressed out!).

Designing the introduction with ethics in mind

While this rule doesn't apply to little polls that you see on the Internet and in magazines, serious surveys need to provide information pertaining to important ethical issues. First, they should include what pollsters call a *cover letter* — an introduction that explains the purpose of the survey, what will be done with the data, whether the information the respondent supplies will be confidential or anonymous (see the sidebar "Anonymity versus confidentiality" later in this chapter), and that the person's participation is appreciated but not required. The cover letter should also provide the researcher's contact information for respondents to use if they have questions or concerns.

If the survey is done by any institution or group that is federally regulated, such as a university, research institute, or a hospital, the survey has to be approved in advance by a committee designated to review, regulate, and/or monitor the research to make sure it's ethical, scientific, and follows regulations. Such committees are called institutional review boards (IRBs), independent ethics committees (IECs), or ethical review boards (ERBs). The survey cover letter should explain who has approved the research. If you don't see such information, ask.

Formulating the questions

After the purpose, type, timing, and ethical issues of the survey have been addressed, the next step is to formulate the questions. The way that the questions are asked can make a huge difference in the quality of the data that will be collected. One of the single most common sources of bias in surveys is the wording of the questions. Research shows that the wording of the questions can directly affect the outcome of a survey. *Leading questions,* also called *misleading questions,* are designed to favor a certain response over another. They can greatly affect how people answer the questions, and their responses may not accurately reflect how they truly feel about an issue.

For example, here are two ways that I've seen survey questions worded about a proposed school bond issue (both of which are leading questions):

> *Don't you agree that a tiny percentage increase in sales tax is a worthwhile investment in improving the quality of the education of our children?*

> *Don't you think we should stop increasing the burden on the taxpayers and stop asking for yet another sales tax hike to fund the wasteful school system?*

From the wording of each of these leading questions, you can easily see how the pollsters want you to respond. To make matters worse, neither question tells you exactly how much of a tax increase is being proposed, which is also misleading.

The best way to word a question is in a neutral way, giving the reader the necessary information required to make an informed decision. For example, the tax issue question is better worded this way:

The school district is proposing a 0.01% increase in sales tax to provide funds for a new high school to be built in the district. What's your opinion on the proposed sales tax? (Possible responses: strongly in favor, in favor, neutral, against, strongly against.)

If the purpose of a survey is purely to collect information rather than influence or persuade the respondent, the questions should be worded in a neutral and informative way in order to minimize bias. The best way to assess the neutrality of a question is to ask yourself whether you can tell how the person wants you to respond. If the answer is yes, that question is a leading question and can give misleading results.

If the results of a survey are important to you, ask the researcher for a copy of the questions used on the survey so you can assess the quality of the questions. When conducting your own survey, have others check the questions to verify that the wording is neutral and informative.

Selecting the sample

After the survey has been designed, the next step is to select people to participate in the survey. Because typically you don't have time or money to conduct a census (a survey of the entire target population), you need to select a subset of the population, called a *sample*. How this sample is selected can make all the difference in terms of the accuracy and the quality of the results.

Three criteria are important in selecting a good sample, as you find out in the following sections.

A good sample represents the target population

To represent the target population, the sample must be selected from the target population, the whole target population, and nothing but the target population. Suppose you want to find out how many hours of TV Americans watch in a day, on average. Asking students in a dorm at a local university to record their TV viewing habits isn't going to cut it. Students represent only a portion of the target population.

Unfortunately, many people who conduct surveys don't take the time or spend the money to select a representative sample of people to participate in the study, and they end up with biased survey results. When presented with survey results, find out how the sample was selected before examining the results of the survey and see how well they match the target population.

A good sample is selected randomly

A *random* sample is one in which every possible sample (of the same size) has an equal chance of being selected from the target population. The easiest example to visualize here is that of a hat (or bucket) containing individual slips of paper, each with the name of a person written on it; if the slips are thoroughly mixed before each slip of paper is drawn out, the result will be a random sample of the target population (in this case, the population of people whose names are in the hat). A random sample eliminates bias in the sampling process.

Reputable polling organizations, such as the Gallup Organization, use a random digit-dialing procedure to telephone the members of their sample. Of course, this excludes people without telephones, but because most American households today have at least one telephone, the bias involved in excluding people without telephones is relatively small.

Beware of surveys that have a large but not randomly selected sample. Internet surveys are the biggest culprit. Someone can say that 50,000 people logged on to a Web site to answer a survey, and that means the person posting this site has gotten a lot of data. But the information is biased; research shows that people who respond to surveys tend to have stronger opinions than those that don't respond. And if they didn't even select the participants randomly to start with, imagine how strong (and biased) the respondents' opinions would be. If the survey designer sampled fewer people but did so randomly, the survey results would be more accurate.

A good sample is large enough for the results to be accurate

If you have a large sample size, and if the sample is representative of the target population and is selected at random, you can count on that information being pretty accurate. *How* accurate depends on the sample size, but the bigger the sample size, the more accurate the information will be (as long as that information is good information). The accuracy of most survey questions is measured in terms of a percentage. This percentage is called the *margin of error,* and it represents how much the researcher expects the results to vary if she were to repeat the survey many times using different samples of the same size. Read more about this in Chapter 12.

A quick and dirty formula to estimate the minimum amount of accuracy of a survey involving categorical data (such as gender or political affiliation) is to take 1 divided by the square root of the sample size. For example, a survey of 1,000 (randomly selected) people is accurate to within ±0.032, or 3.2 percentage points. (See Chapter 12 for the exact formula for calculating the accuracy of a survey.) In cases where not everyone responded, you should replace the sample size with the number of respondents (see the "Following up, following up, and following up" section later in this chapter). Remember, these quick-and-dirty estimates of accuracy are conservative; using the precise formulas gives you accuracy rates that are often much better than these. (See Chapter 13 for details.)

With large populations (in the thousands, say) it's the size of the sample, not the size of the population, that matters. For example, if you randomly sample 1,000 individuals from a large population, your accuracy level is estimated to be within 3.2 percentage points, no matter whether you sample from a small town of 10,000 people, a state of 1,000,000 people, or all of the United States. That fact was one of the things that blew my mind about statistics when I first learned it, and it still does today — it's amazing how accurate you can get with such a comparatively small sample size.

However, with small populations, you have to apply different methods to determine accuracy and sample size. A sample of 10 out of a population of 100 takes a much larger piece out of the pie than a sample of 10 out of 10,000 does, for example. More advanced methods involving a finite population correction handle issues that come up with small populations.

Carrying out a survey

The survey has been designed, and the participants have been selected. Now you have to go about the process of carrying out the survey, which is another important step — one where lots of mistakes and biases can occur.

Collecting the data

During the survey itself, the participants can have problems understanding the questions, they may give answers that aren't among the choices (in the case of a multiple choice question), or they may decide to give answers that are inaccurate or blatantly false; the latter is called *response bias*. (As an example of response bias, think about the difficulties involved in getting people to tell the truth about whether they've cheated on their income-tax forms.)

Some of the potential problems with the data-collection process can be minimized or avoided with careful training of the personnel who carry out the survey. With proper training, any issues that arise during the survey are resolved in a consistent and clear way, and fewer errors are made in recording the data. Problems with confusing questions or incomplete choices for answers can be resolved by conducting a pilot study on a few participants prior to the actual survey and then, based on their feedback, fixing any problems with the questions.

Personnel can also be trained to create an environment in which each respondent feels safe enough to tell the truth; ensuring that privacy will be protected also helps encourage more people to respond. To minimize interviewer bias, the interviewers must follow a script that's the same for each subject.

Anonymity versus confidentiality

If you were to conduct a survey to determine the extent of personal e-mail use at work, the response rate would probably be an issue, because many people are reluctant to discuss their use of personal e-mail in the workplace, or at least to do so truthfully. You could try to encourage people to respond by letting them know that their privacy would be protected during and after the survey.

When you report the results of a survey, you generally don't tie the information collected to the names of the respondents, because doing so would violate the privacy of the respondents. You've probably heard the terms *anonymous* and *confidential* before, but what you may not

realize is that these two words are completely different in terms of privacy issues. Keeping results *confidential* means that I could tie your information to your name in my report, but I promise that I won't do that. Keeping results *anonymous* means that I have no way of tying your information to your name in my report, even if I wanted to.

If you're asked to participate in a survey, be sure you're clear about what the researchers plan to do with your responses and whether or not your name can be tied to the survey. (Good surveys always make this issue very clear for you.) Then make a decision as to whether you still want to participate.

 Beware of conflicts of interest that come up with misleading surveys. For example, if you are being asked about the quality of your service by the person who gave you the service, you may not want to respond truthfully. Or, if your physical therapist gives you an "anonymous" feedback survey on your last day and tells you to give it to her when you're done, the survey may have issues of bias.

Following up, following up, and following up

Anyone who has ever thrown away a survey or refused to "answer a few questions" over the phone knows that getting people to participate in a survey isn't easy. If the researcher wants to minimize bias, the best way to handle it is to get as many folks to respond as possible by following up, one, two, or even three times. Offer dollar bills, coupons, self-addressed stamped return envelopes, chances to win prizes, and so on. Every little bit helps.

If only those folks who feel very strongly respond to a survey, that means that only their opinions will count, because the other people who didn't really care about the issue didn't respond, and their "I don't care" vote didn't get counted. Or maybe they did care, but they just didn't take the time to tell anyone. Either way, their vote doesn't count.

For example, suppose 1,000 people are given a survey about whether the park rules should be changed to allow dogs without leashes. Most likely, the respondents would be those who strongly agree or disagree with the proposed rules. Suppose only 200 people responded — 100 against and 100 for the issue. That would mean that 800 opinions weren't counted. Suppose none of those 800 people really cared about the issue either way. If you could count their opinions, the results would be $800 \div 1,000 = 80\%$ "no opinion," $100 \div 1,000 = 10\%$ in favor of the new rules, and $100 \div 1,000 = 10\%$ against the new rules. But without the votes of the 800 non-respondents, the researchers would report, "Of the people who responded, 50% were in favor of the new rules and 50% were against them." This gives the impression of a very different (and a very biased) result from the one you would've gotten if all 1,000 people had responded.

The *response rate* of a survey is a ratio found by taking the number of respondents divided by the number of people who were originally asked to participate. You of course want to have the highest response rate you can get with your survey; but how high is high enough to be minimizing bias? The purest of the pure statisticians feel that a good response rate is anything over 70%, but I think we need to be a little more realistic. Today's fast-paced society is saturated with surveys; many if not most response rates fall far short of 70%. In fact, response rates for today's surveys are more likely to be in the 20% to 30% range, unless the survey is conducted by a professional polling organization such as Gallup or you are being offered a new car just for filling one out.

Look for the response rate when examining survey results. If the response rate is too low (much less than 50%) the results are likely to be biased and should be taken with a grain of salt, or even ignored.

Don't be fooled by a survey that claims to have a large number of respondents but actually has a low response rate; in this case, many people may have responded, but many more were asked and didn't respond.

Note that statistical formulas at this level (including the formulas in this book) assume that your sample size is equal to the number of respondents, so statisticians want you to know how important it is to follow up with people and not end up with biased data due to non-response. However, in reality, statisticians know that you can't always get everyone to respond, no matter how hard you try; indeed, even the U.S. Census doesn't have a 100% response rate. One way statisticians combat the non-response problem after the data have been collected is to break down the data to see how well it matches the target population. If it's a fairly good match, they can rest easier on the bias issue.

So which number do you put in for n in all those statistical formulas you use so often (such as the sample mean in Chapter 5)? You can't use the intended sample size (the number of people contacted). You have to use the final sample size (the number of people who responded). In the media you most often see only the number of respondents reported, but you also need the response rate (or the total number of respondents) to be able to critically evaluate the results.

Regarding the quality of results, selecting a smaller initial sample size and following them up more aggressively is a much better approach than selecting a larger group of potential respondents and having a low response rate, because of the bias introduced by non-response.

Interpreting results and finding problems

The purpose of a survey is to gain information about your target population; this information can include opinions, demographic information, or lifestyles and behaviors. If the survey has been designed and conducted in a fair and accurate manner with the goals of the survey in mind, the data should provide good information as to what's happening with the target population (within the stated margin of error; see Chapter 12). The next steps are to organize the data to get a clear picture of what's happening; to analyze the data to look for links, differences, or other relationships of interest; and then to draw conclusions based on the results.

Organizing and analyzing

After a survey has been completed, the next step is to organize and analyze the data (in other words, crunch some numbers and make some graphs). Many different types of data displays and summary statistics can be created and calculated from survey data, depending on the type of information that was collected. (Numerical data, such as income, have different characteristics and are usually presented differently than categorical data, such as gender.) For more information on how data can be organized and summarized, see Chapters 5 through 7. Depending on the research question, different types of analyses can be performed on the data, including coming up with population estimates, testing a hypothesis about the population, or looking for relationships, to name a few. See Chapters 13, 14, 15, 18, and 19 for more on each of these analyses, respectively.

Watch for misleading graphs and statistics. Not all survey data are organized and analyzed fairly and correctly. See Chapter 3 for more about how statistics can go wrong.

Drawing conclusions

The conclusions are the best part of any survey — they're why the researchers do all of the work in the first place. If the survey was designed and carried out properly — the sample was selected carefully and the data were organized and summarized correctly — the results should fairly and accurately represent the reality of the target population. But, of course, not all surveys are done right. And even if a survey is done correctly, researchers can misinterpret or overinterpret results so that they say more than they really should.

You know the saying "Seeing is believing"? Some researchers are guilty of the converse, which is "Believing is seeing." In other words, they claim to see what they want to believe about the results. All the more reason for you to know where the line is drawn between reasonable conclusions and misleading results, and to realize when others have crossed that line.

Here are some common errors made in drawing conclusions from surveys:

✔ Making projections to a larger population than the study actually represents

✔ Claiming a difference exists between two groups when a difference isn't really there (see Chapter 15)

✔ Saying, "these results aren't scientific, but . . . ," and then going on to present the results as if they are scientific

To avoid common errors made when drawing conclusions, do the following:

1. **Check whether the sample was selected properly and that the conclusions don't go beyond the population presented by that sample.**

2. **Look for any disclaimers about the survey *before* reading the results.**

   That way, if the results aren't based on a scientific survey (an accurate and unbiased survey), you'll be less likely to be influenced by the results you're reading. You can judge for yourself whether the survey results are credible.

3. **Be on the lookout for statistically incorrect conclusions.**

   If someone reports a difference between two groups in terms of survey results, be sure that the difference is larger than the reported margin of error. If the difference is within the margin of error, you should expect the sample results to vary by that much just by chance, and the so-called "difference" can't really be generalized to the entire population. (See Chapter 14 for more on this.)

Know the limitations of any survey and be wary of any information coming from surveys in which those limitations aren't respected. A bad survey is cheap and easy to do, but you get what you pay for. But don't let big expensive surveys fool you either — they can be riddled with bias as well! Before looking at the results of any survey, investigate how it was designed and conducted, using the criteria and tips in this chapter, so you can judge the quality of the results and express yourself confidently and correctly about what is wrong.

Chapter 17

Experiments: Medical Breakthroughs or Misleading Results?

Medical breakthroughs seem to come and go quickly. One day you hear about a promising new treatment for a disease, only to find out later that the drug didn't live up to expectations in the last stage of testing. Pharmaceutical companies bombard TV viewers with commercials for pills, sending millions of people to their doctors clamoring for the latest and greatest cures for their ills, sometimes without even knowing what the drugs are for. Anyone can search the Internet for details about any type of ailment, disease, or symptom and come up with tons of information and advice. But how much can you really believe? And how do you decide which options are best for you if you get sick, need surgery, or have an emergency?

In this chapter, you go behind the scenes of experiments, the driving force of medical studies and other investigations in which comparisons are made — comparisons that test, for example, which building materials are best, which soft drink teens prefer, and so on. You find out the difference between experiments and observational studies and discover what experiments can do for you, how they're supposed to be done, how they can go wrong, and how you can spot misleading results. With so many headlines, sound bites, and pieces of "expert advice" coming at you from all directions, you need to use all your critical thinking skills to evaluate the sometimes-conflicting information you're presented with on a regular basis.

Boiling Down the Basics of Studies

Although many different types of studies exist, you can basically boil them down to two types: experiments and observational studies. This section examines what exactly makes experiments different from other studies. But before I dive in to the details, I need to lay some jargon on you.

Looking at the lingo of studies

To understand studies, you need to find out what their commonly used terms mean:

- **Subjects:** Individuals participating in the study.

- **Observational study:** A study in which the researcher merely observes the subjects and records the information. No intervention takes place, no changes are introduced, and no restrictions or controls are imposed.

- **Experiment:** This study doesn't simply observe subjects in their natural state; it deliberately applies treatments to them in a controlled situation and studies their effects on the outcome.

- **Response:** The response is the variable whose outcome is the million dollar question; it's the variable whose outcome is of interest. For example, if researchers want to know what happens to your blood pressure when you take a large amount of Ibuprofen each day, the response variable is blood pressure.

- **Factor:** A factor is the variable whose effect on the response is being studied. For example, if you want to know whether a particular drug increases blood pressure, your factor is the amount of the drug taken. If you want to know which weight loss program is most effective, your factor would be the type of weight loss program used.

  You can have more than one factor in a study; however, in this book I stick with discussing one factor only. For the analysis of two-factor studies, including the use of Analysis of Variance (ANOVA) and multiple comparisons to compare treatment combinations, you can check out my book *Statistics II For Dummies,* also published by Wiley.

- **Level:** A level is one possible outcome of a factor. Each factor has a certain number of levels. In the weight loss example, the factor is the type of weight loss program and the levels would be the specific programs studied (for example Weight Watchers, South Beach, or the famous Potato Diet). Levels need not be ascending in any way; however, in a study like the drug example, the levels would be the various dosages taken each day, in increasing amounts.

✔ **Treatment:** A treatment is a combination of the levels of the factors being studied. If you only have one factor, the levels and the treatments are the same thing. If you have more than one factor, each combination of levels of the factors is called a treatment.

For example, if you want to study the effects of the type of weight loss program and the amount of water consumed daily, you have two factors: 1) the type of program, with 3 levels (Weight Watchers, South Beach, Potato Diet); and 2) the amount of water consumed, with, say, 3 levels (24, 48, and 64 ounces per day). In this case, there are $3 * 3 = 9$ treatments: Weight Watchers and 24 ounces of water per day; Weight Watchers and 48 ounces of water per day, . . . all the way up to the famous Potato Diet and 64 ounces of water per day. Each subject is assigned to one treatment. (With my luck, I'd get that last treatment.)

✔ **Cause and effect:** A factor and a response have a cause-and-effect relationship if a change in the factor results in a direct change in the response (for example, increasing calorie intake causes weight gain).

In the following sections, you see the differences between observational studies and experiments, when each is used, and what their strengths and/or weaknesses may be.

Observing observational studies

Just like with tools, you want to find the right type of study for the right job. In certain situations, observational studies are the optimal way to go. The most common observational studies are *polls* and *surveys* (see Chapter 16). When the goal is simply to find out what people think and to collect some demographic information (such as gender, age, income, and so on), surveys and polls can't be beat, as long as they're designed and conducted correctly.

In other situations, especially those looking for cause-and-effect relationships, observational studies aren't optimal. For example, suppose you took a couple of vitamin C pills last week; is that what helped you avoid getting that cold that's going around the office? Maybe the extra sleep you got recently or the extra hand-washing you've been doing helped you ward off the cold. Or maybe you just got lucky this time. With so many variables in the mix, how can you tell which one had an influence on the outcome of your not getting a cold? An experiment that takes these other variables into account is the way to go.

When looking at the results of any study, first determine what the purpose of the study was and whether the type of study fits the purpose. For example, if an observational study was done instead of an experiment to establish a cause-and-effect relationship, any conclusions that are drawn should be carefully scrutinized.

Examining experiments

The object of an experiment is to see if the response changes as a result of the factor you are studying; that is, you are looking for cause and effect. For example, does taking Ibuprofen cause blood pressure to increase? If so, by how much? But because results will vary with any experiment, you want to know that your results have a high chance of being repeatable if you found something interesting happening. That is, you want to know that your results were unlikely to be due to chance; statisticians call such results *statistically significant*. That's the objective of any study, observational, or experimental.

A good experiment is conducted by creating a very controlled environment — so controlled that the researcher can pinpoint whether a certain factor or combination of factors causes a change in the response variable, and if so, the extent to which that factor (or combination of factors) influences the response. For example, to gain government approval for a proposed blood pressure drug, pharmaceutical researchers set up experiments to determine whether that drug helps lower blood pressure, what dosage level is most appropriate for each different population of patients, what side effects (if any) occur, and to what extent those side effects occur in each population.

Designing a Good Experiment

How an experiment is designed can mean the difference between good results and garbage. Because most researchers are going to write the most glowing press releases that they can about their experiments, you have to be able to sort through the hype to determine whether to believe the results you're being told. To decide whether an experiment is credible, check to see if it meets *all* the following criteria for a good experiment. A good experiment:

- ✔ **Makes comparisons**
- ✔ **Includes a large enough sample size so that the results are accurate**
- ✔ **Chooses subjects that most accurately represent the target population**
- ✔ **Assigns subjects randomly to the treatment group(s) and the control group**
- ✔ **Controls for possible confounding variables**
- ✔ **Is ethical**
- ✔ **Collects good data**
- ✔ **Applies the proper data analysis**
- ✔ **Makes appropriate conclusions**

In this section, each of these criteria is explained and illustrated with examples.

Designing the experiment to make comparisons

Every experiment has to make bonafide comparisons to be credible. This seems to go without saying, but researchers often are so gung-ho to prove their results that they forget (or just don't bother) to show that their factor, and not some other factor(s), including random chance, was the actual cause for any differences found in the response.

For example, suppose a researcher is convinced that taking vitamin C prevents colds, and she assigns subjects to take one vitamin C pill per day and follows them for 6 months. Suppose the subjects get very few colds during that time. Can she attribute these results to the vitamin C and nothing else? No; there's no way of knowing whether the subjects would have been just as healthy without the vitamin C, due to some other factor(s), or just by chance. There's nothing to compare the results to.

To tease out the real effect (if any) that your factor has on the response, you need a baseline to compare the results to. This baseline is called the *control*. Different methods exist for creating a control in an experiment; depending on the situation, one method typically rises to the top as being the most appropriate. Three common methods for including control are to administer: 1) a fake treatment; 2) a standard treatment; or 3) no treatment. The following sections describe each method.

When examining the results of an experiment, make sure the researchers established a baseline by creating a control group. Without a control group, you have nothing to compare the results to, and you never know whether the treatment being applied was the real cause of any differences found in the response.

Fake treatments — the placebo effect

A fake treatment (also called a *placebo*) is not distinguishable from a "real" treatment by the subject. For example, when drugs are administered, a subject assigned to the placebo will receive a fake pill that looks and tastes exactly like a real pill; it's just filled with an inert substance like sugar instead of the actual drug. A placebo establishes a baseline measure for what responses would have taken place anyway, in lieu of any treatment (this would have helped the vitamin C study mentioned under "Designing the experiment to make comparisons"). But a fake treatment also takes into account what researchers call the *placebo effect,* a response that people have (or think they're having) because they know they're getting some type of "treatment" (even if that treatment is a fake treatment, such as sugar pills).

Pharmaceutical companies are required to account for the placebo effect when examining both the positive and negative effects of a drug. When you see an ad for a drug in a magazine, you see the positive results of the drug standing out in big, bright, happy, colorful visuals. Then look at the back of the page and you see it's entirely filled in black with words written in 3-point font. Embedded somewhere on that page, you can find one or more tiny tables that show the number and nature of side effects reported by each *treatment group* (subjects who received an actual treatment) as well as the *control group* (subjects who were administered a placebo).

If the control group is on a placebo, you may expect the subjects not to report any side effects, but you would be wrong. If you are taking a pill, you know it could be an actual drug, and you are being asked whether or not you're experiencing side effects, you might be surprised at what your response would be.

If you don't take the placebo effect into account, you have to believe that any side effects (or positive results) reported are actually due to the drug. This gives an artificially high number of reported side effects because at least some of those reports are likely due to the placebo effect and not to the drug itself. If you have a control group to compare with, you can subtract the percentage of people in the control group who reported the side effects from the percentage of people in the treatment group that reported the side effects, and examine the magnitude of the numbers that remain. You're in essence looking at the net number of reported side effects due to the drug, rather than the gross number of side effects, some of which are due to the placebo effect.

The placebo effect has been shown to be real. If you want to be fair about examining the reported side effects (or positive reactions) of a treatment, you have to also take into account the side effects (or positive reactions) that the control group reports — those reactions that are due to the placebo effect only.

Standard treatments

In some situations, such as when the subjects have very serious diseases, offering a fake treatment as an option may be unethical. One famous example of a breech in ethics occurred in 1997. The U.S. government was harshly criticized for financing an HIV study that examined new dosage levels of AZT, a drug known at that time to cut the risk of HIV transmission from pregnant mothers to their babies by two-thirds. This particular study, in which 12,000 pregnant women with HIV in Africa, Thailand, and the Dominican Republic participated, had a deadly design. Researchers gave half of the women various dosages of AZT, but the other half of the women received sugar pills. Of course, had the U.S. government realized that a placebo was being given to half of the subjects, it wouldn't have supported the HIV study. It's not ethical to give a fake treatment to anyone with a deadly disease for which a standard treatment is available (in this case, the standard dosage of AZT).

When ethical reasons bar the use of fake treatments, the new treatment is compared to at least one existing or standard treatment that is known to be an effective treatment. After researchers have enough data to see that one of the treatments is working better than the other, they generally stop the experiment and put everyone on the better treatment; again, for ethical reasons.

No treatment

"No treatment" means the researcher can't help but tell which group the subject is in, due to the nature of the experiment. The subjects in this case aren't receiving any type of intervention in terms of their behavior, but they still serve as a control, establishing a baseline of data to compare their results with those in the treatment group(s). For example, if you want to determine whether speed walking around the block ten times a day lowers a person's resting heart rate after six months, the subjects in your control group know they aren't going to be speed walking — obviously you can't do fake speed walking (although faking exercising and still reaping the benefits would be great, wouldn't it?).

In situations where the control group receives no treatment, you still make sure the groups of subjects (speed walkers versus non–speed walkers) are similar in as many ways as possible, and that the other criteria for a good experiment are being met. (See "Designing a Good Experiment" for the list of criteria.)

Selecting the sample size

The size of a (good) sample greatly affects the accuracy of the results. The larger the sample size, the more accurate the results, and the more powerful the statistical tests (in terms of being able to detect real results when they exist). In this section, I hit the highlights; Chapter 14 has the details.

The word *sample* is often attributed to surveys where a random sample is selected from the target population (see Chapter 16). However, in the setting of experiments, a sample means the group of subjects who have volunteered to participate.

Limiting small samples to small conclusions

You may be surprised at the number of research headlines that have been made regarding large populations that were based on very small samples. Such headlines can be of concern to statisticians, who know that detecting true statistically significant results in a large population using a small sample is difficult because small data sets have more variability from sample to sample (see Chapter 12). When sample sizes are small and big conclusions have been made by the researcher, either the researchers didn't use the right hypothesis test to analyze their data (for example, using the Z-distribution rather than the t-distribution; see Chapter 10) or the difference was so large that it would be very difficult to miss. The latter isn't always the case, however.

Be wary of research conclusions that find significant results based on small sample sizes (especially for experiments involving many treatments but only a few subjects assigned to each treatment). Statisticians want to see at least five subjects per treatment, but (much) more is (much) better. You do need to be aware of some of the limitations of experiments such as cost, time, as well as ethical issues, and realize that the number of subjects for experiments is often smaller than the number of participants in a survey.

If the results are important to you, ask for a copy of the research report and look to see what type of analysis was done on the data. Also look at the sample of subjects to see whether this sample truly represents the population about which the researchers are drawing conclusions.

Defining sample size

When asking questions about *sample size*, be specific about what you mean by the term. For example, you can ask how many subjects were selected to participate and also ask for the number who actually completed the experiment; these two numbers can be very different. Make sure the researchers can explain any situations in which the research subjects decided to drop out or were unable (for some reason) to finish the experiment.

For example, an article in *The New York Times* titled "Marijuana Is Called an Effective Relief in Cancer Therapy" says in the opening paragraph that marijuana is "far more effective" than any other drug in relieving the side effects of chemotherapy. When you get into the details, you find out that the results are based on only 29 patients (15 on the treatment, 14 on a placebo). Then you find out that only 12 of the 15 patients in the treatment group actually completed the study. What happened to the other three subjects?

Sometimes researchers draw their conclusions based on only those subjects who completed the study. This can be misleading, because the data don't include information about those who dropped out (and why), which may be leading to biased data. For a discussion of the sample size you need to achieve a certain level of accuracy, see Chapter 13.

Accuracy isn't the only issue in terms of having "good" data. You still need to worry about eliminating bias by selecting a random sample (see Chapter 16 for more on how random samples are taken).

Choosing the subjects

The first step in carrying out an experiment is selecting the subjects (participants). Although researchers would like their subjects to be selected randomly from their respective populations, in most cases, this just isn't appropriate. For example, suppose a group of eye researchers wants to test out a new laser surgery on nearsighted people. They need a random sample of subjects, so they randomly select various eye doctors from across the country and randomly select nearsighted patients from these doctors' files. They call up each person selected and say, "We're experimenting with a new laser surgery technique for nearsightedness, and you've been selected at random to participate in our study. When can you come in for the surgery?" Something tells me that this approach wouldn't go over very well with many people receiving the call (although some would probably jump at the chance, especially if they didn't have to pay for the procedure).

The point is that getting a truly random sample of people to participate in an experiment is generally more difficult than getting a random sample of folks to participate in a survey. However, statisticians can build techniques into the design of an experiment to help minimize the potential bias that can occur.

Making random assignments

One way to minimize bias in an experiment is to introduce some randomness. After the sample has been decided on, the subjects are randomly divided into treatment and control groups. The treatment groups receive the various treatments being studied, and the control group receives the current (or standard) treatment, no treatment, or a placebo. (See the section "Designing the experiment to make comparisons" earlier in this chapter.)

Making random assignments of subjects to treatments is an extremely critical step toward minimizing bias in an experiment. Suppose a researcher wants to determine the effects of exercise on heart rate. The subjects in his treatment group run 5 miles and have their heart rates measured before and after the run. The subjects in his control group sit on the couch the whole time and watch reruns of old TV shows. Which group would you rather be in? Some health nuts out there would no doubt volunteer for the treatment group. If you're not crazy about the idea of running five miles, you may opt for the easy way out and volunteer to be a couch potato. (Or maybe you hate to watch old reruns so much that you'd run five miles to avoid that.)

Finding volunteers

To find subjects for their experiments, researchers often advertise for volunteers and offer them incentives such as money, free treatments, or follow-up care for their participation. Medical research on humans is complicated and difficult, but it's necessary in order to really know whether a treatment works, how well it works, what the dosage should be, and what the side effects are. In order to prescribe the right treatments in the right amounts in real-life situations, doctors and patients depend on these studies being representative of the general population. In order to recruit such representative subjects, researchers have to do a broad advertisement campaign and select enough participants with enough different characteristics to represent a cross section of the populations of folks who will be prescribed these treatments in the future.

What impact would this selective volunteering have on the results of the study? If only the health nuts (who probably already have excellent heart rates) volunteer to be in the treatment group, the researcher will be looking only at the effect of the treatment (running five miles) on very healthy and active people. He won't see the effect that running five miles has on the heart rates of couch potatoes. This non-random assignment of subjects to the treatment and control groups could have a huge impact on the conclusions he draws from this study.

To avoid major bias in the results of an experiment, subjects must be randomly assigned to treatments by a third party and not be allowed to choose which group they will be in. The goal of random assignment is to create homogenous groups; any unusual characteristics or biases have an equal chance of appearing in any of the groups. Keep this in mind when you evaluate the results of an experiment.

Controlling for confounding variables

Suppose you're participating in a research study that looks at factors influencing whether you catch a cold. If a researcher records only whether you got a cold after a certain period of time and asks questions about your behavior (how many times per day you washed your hands, how many hours of sleep you get each night, and so on), the researcher is conducting an observational study. The problem with this type of observational study is that without controlling for other factors that may have had an influence and without regulating which action you were taking when, the researcher won't be able to single out exactly which of your actions (if any) actually impacted the outcome.

The biggest limitation of observational studies is that they can't really show true cause-and-effect relationships, due to what statisticians call confounding variables. A *confounding variable* is a variable or factor that was not controlled for in the study but can have an influence on the results.

For example, one news headline boasted, "Study links older mothers, long life." The opening paragraph said that women who have a first baby after age 40 have a much better chance of living to be 100, compared to women who have a first baby at an earlier age. When you get into the details of the study (done in 1996) you find out, first of all, that it was based on 78 women in suburban Boston who were born in 1896 and had lived to be at least 100, compared to 54 women who were also born in 1896 but died in 1969 (the earliest year the researchers could get computerized death records). This so-called "control group" lived to be exactly 73, no more and no less. Of the women who lived to be at least 100 years of age, 19% had given birth after age 40, whereas only 5.5% of the women who died at age 73 had given birth after age 40.

I have a real problem with these conclusions. What about the fact that the "control group" was based only on mothers who died in 1969 at age 73? What about all the other mothers who died *before* age 73, or who died between the ages of 73 and 100? What about other variables that may affect both mothers' ages at the births of their children and longer life spans — variables such as financial status, marital stability, or other socioeconomic factors? The women in this study were in their thirties during the Depression; this may have influenced both their life span and if or when they had children.

How do researchers handle confounding variables? They control for them as best they can, for as many of them as they can anticipate, trying to minimize their possible effect on the response. In experiments involving human subjects, researchers have to battle many confounding variables.

For example, in a study trying to determine the effect of different types and volumes of music on the amount of time grocery shoppers spend in the store (yes, they do think about that), researchers have to anticipate as many possible confounding variables ahead of time and then control for them. What other factors besides volume and type of music may influence the amount of time you spend in a grocery store? I can think of several factors: gender, age, time of day, whether you have children with you, how much money you have, the day of the week, how clean and inviting the store is, how nice the employees are, and (most importantly) what your motive is — are you shopping for the whole week, or are you just running in to grab a candy bar?

How can researchers begin to control for so many possible confounding factors? Some of them can be controlled for in the design of the study, such as the time of the day, day of the week, and reason for shopping. But other factors (such as the perception of the store environment) depend totally on the individual in the study. The ultimate form of control for those person-specific confounding variables is to use pairs of people that are matched according to important variables, or to just use the same person twice: once with the treatment and once without. This type of experiment is called a *matched-pairs design*. (See Chapter 15 for more on this.)

Before believing any medical headlines (or any headlines with statistics, for that matter), look to see how the study was conducted. Observational studies can't control for confounding variables, so their results are not as statistically meaningful (no matter what the statistics say) as the results of a well-designed experiment are. In cases where an experiment can't be done (after all, no one can force you to have a baby after or before age 40), make sure the observational study is based on a large enough sample that represents a cross-section of the population. And think about possible confounding variables that may affect the conclusions being drawn.

Respecting ethical issues

The trouble with experiments is that some experimental designs are not ethical. You can't force research subjects to smoke in order to see whether they get lung cancer, for example — you can only look at people who have lung cancer and work backward to see what *factors* (variables being studied) may have caused the disease. But because you can't control for the various factors you're interested in — or for any other variables, for that matter — singling out any one particular cause becomes difficult with observational studies. That's why so much evidence was needed to show that smoking causes lung cancer, and why the tobacco companies only recently had to pay huge penalties to victims.

Although the causes of cancer and other diseases can't be determined ethically by conducting experiments on humans, new treatments for cancer can be (and are) tested using experiments. Medical studies that involve experiments are called *clinical trials*. The U.S. government has a registry of federally and privately supported clinical trials conducted in the United States and around the world; it also has information available on who may participate in various clinical trials. Check out www.clinicaltrials.gov for more information.

Serious experiments (such as those funded by and/or regulated by the U.S. government) must pass a huge series of tests that can take years to carry out. The approval of a new drug, for example, goes through a very lengthy, comprehensive, and detailed process regulated and monitored by the FDA (Federal Drug Administration). One reason the cost of prescription drugs is so high is the massive amount of time and money needed to conduct research and development of new drugs, most of which fail to pass the tests and have to be scrapped.

Any experiments involving human subjects are also regulated by the federal government and have to gain approval by a committee created for the purpose of protecting "the rights and welfare of the participants." The committees set up for different organizations have different names (such as Institutional Review Board [IRB], Independent Ethics Committee [IEC], or Ethical Review Board [ERB], to name a few) but they all serve the same purpose. Research conducted on animals is more nebulous in terms of regulations and continues to be a topic of much debate and controversy in the U.S. and around the world.

Surveys, polls, and other observational studies are fine if you want to know people's opinions, examine their lifestyles without intervention, or examine some demographic variables. If you want to try to determine the cause of a certain outcome or behavior (that is, a reason why something happened), an experiment is a much better way to go. If an experiment isn't possible because of ethics concerns (or because of expense or other reasons), a large body of observational studies examining many different factors and coming up with similar conclusions is the next best thing. (See Chapter 18 for more about cause-and-effect relationships.)

Collecting good data

What constitutes "good" data? Statisticians use three criteria for evaluating data quality; each of the criteria really relates most strongly to the quality of the measurement instrument that's used in the process of collecting the data. To decide whether you're looking at good data from a study, look for these characteristics:

- ✔ **The data are reliable — you can get repeatable results with subsequent measurements.** Many bathroom scales give unreliable data. You get on the scale, and it gives you one number. You don't believe the number, so you get off, get back on, and get a different number. (If the second number is lower, you'll most likely quit at this point; if not, you may continue getting on and off until you see a number you like.) Or you can do what some researchers do: Take three measurements, find the average, and use that; at least this will improve the reliability a bit.

  Unreliable data come from unreliable measurement instruments or unreliable data collection methods. Errors can go beyond the actual scales to more intangible measurement instruments, like survey questions, which can give unreliable results if they're written in an ambiguous way (see Chapter 16).

  Find out how the data were collected when examining the results of a study. If the measurements are unreliable, the data could be inaccurate.

- ✔ **The data are valid — they measure what they're supposed to measure.** Checking the validity of data requires you to step back and look at the big picture. You have to ask the question: Do these data measure what they should be measuring? Or should the researchers have been collecting altogether different data? The appropriateness of the measurement instrument used is important. For example, many educators say that a student's transcript is not a valid measure of their ability to perform well in college. Alternatives include a more holistic approach, taking into account not only grades, but adding weight to elements such as service, creativity, social involvement, extracurricular activities, and the like.

  Before accepting the results of an experiment, find out what data were measured and how they were measured. Be sure the researchers are collecting valid data that are appropriate for the goals of the study.

✔ **The data are unbiased — they contain no systematic errors that either add to or subtract from the true values.** Biased data are data that systematically overmeasure or undermeasure the true result. Bias can occur almost anywhere during the design or implementation of a study. Bias can be caused by a bad measurement instrument (like that bathroom scale that's "always" 5 pounds over), by survey questions that lead participants in a certain way, or by researchers who know what treatment each subject received and who have preconceived expectations.

Bias is probably the number-one problem in collecting good data. However, you can minimize bias with methods similar to those discussed in Chapter 16 for surveys and in the "Making random assignments" section earlier in this chapter, and by making your experiments double-blind whenever possible.

Double-blind means neither the subjects nor the researchers know who got what treatment or who is in the control group. The subjects need to be oblivious to which treatment they're getting so that the researchers can measure the placebo effect. And researchers should be kept in the dark so they don't treat subjects differently by either expecting or not expecting certain responses from certain groups. For example, if a researcher knows you're in the treatment group to study the side effects of a new drug, she may expect you to get sick and therefore may pay more attention to you than if she knew you were in the control group. This can result in biased data and misleading results.

If the researcher knows who got what treatment but the subjects don't know, the study is called a *blind* study (rather than a double-blind study). Blind studies are better than nothing, but double-blind studies are best. In case you're wondering: In a double-blind study, does *anyone* know which treatment was given to which subjects? Relax; typically a third party, such as a lab assistant, does that part.

In some cases the subjects know which group they're in because it's unconcealable — for example, when comparing the benefits of doing yoga versus jogging. However, bias can be reduced by not telling the subjects the precise purpose of the study. This irregular type of plan would have to be reviewed by an institutional review board to make sure it isn't unethical to do; see the earlier section "Respecting ethical issues."

Analyzing the data properly

After the data have been collected, they're put into that mysterious box called the *statistical analysis* for number crunching. The choice of analysis is just as important (in terms of the quality of the results) as any other aspect of a study. A proper analysis should be planned in advance, during the design phase of the experiment. That way, after the data are collected, you won't run into any major problems during the analysis.

Here's the bottom line when selecting the proper analysis: Ask yourself the question, "After the data are analyzed, will I be able to legitimately and correctly answer the question that I set out to answer?" If the answer is "no," then that analysis isn't appropriate.

Some basic types of statistical analyses include *confidence intervals* (used when you're trying to estimate a population value, or the difference between two population values); *hypothesis tests* (used when you want to test a claim about one or two populations, such as the claim that one drug is more effective than another); and *correlation and regression analyses* (used when you want to show if and/or how one quantitative variable can predict or cause changes in another quantitative variable). See Chapters 13, 15, and 18, respectively, for more on each of these types of analyses.

When choosing how you're going to analyze your data, you have to make sure that the data and your analysis will be compatible. For example, if you want to compare a treatment group to a control group in terms of the amount of weight lost on a new (versus an existing) diet program, you need to collect data on how much weight each person lost — not just each person's weight at the end of the study.

Making appropriate conclusions

In my opinion, the biggest mistakes researchers make when drawing conclusions about their studies are the following (discussed in the following sections):

- ✔ Overstating their results
- ✔ Making connections or giving explanations that aren't backed up by the statistics
- ✔ Going beyond the scope of the study in terms of whom the results apply to

Overstating the results

Many times, the headlines in the media overstate actual research results. When you read a headline or otherwise hear about a study, be sure to look further to find out the details of how the study was done and exactly what the conclusions were.

Press releases often overstate results, too. For example, in a recent press release by the National Institute for Drug Abuse, the researchers claimed that use of the street drug Ecstasy was down from the previous year. However, when you look at the actual statistical results in the report, you find that the percentage of teens *from the sample* who said they'd used Ecstasy was lower than those from the previous year, but this difference was not found to be statistically significant when they tried to project it onto the population of *all* teens. This discrepancy means that although fewer teens in the sample used Ecstasy that year, the difference wasn't enough to account for more than chance variability from sample to sample. (See Chapter 14 for more about statistical significance.)

Headlines and leading paragraphs in press releases and news articles often overstate the actual results of a study. Big results, spectacular findings, and major breakthroughs make the news these days, and reporters and others in the media constantly push the envelope in terms of what is and isn't newsworthy. How can you sort out the truth from exaggeration? The best thing to do is to read the fine print.

Taking the results one step beyond the actual data

A study that links having children later in life to longer life spans illustrates another point about research results. Do the results of this observational study mean that having a baby later in life can make you live longer? "No," said the researchers. Their explanation of the results was that having a baby later in life may be due to women having a "slower" biological clock, which presumably would then result in the aging process being slowed down.

My question to these researchers is, "Then why didn't you study *that,* instead of just looking at their ages?" The study didn't include any information that would lead me to conclude that women who had children after age 40 aged at a slower rate than other women, so in my view, the researchers shouldn't make that conclusion. Or the researchers should state clearly that this view is only a theory and requires further study. Based on the data in this study, the researchers' theory seems like a leap of faith (although since I became a new mom at age 41, I'll hope for the best!).

Frequently in a press release or news article, the researcher will give an explanation about *why* he thinks the results of the study turned out the way they did and what implications these results have for society as a whole when the "why" hasn't been studied yet. These explanations may have been in response to a reporter's questions about the research — questions that were later edited out of the story, leaving only the juicy quotes from the researcher. Many of these after-the-fact explanations are no more than theories that have yet to be tested. In such cases, you should be wary of conclusions, explanations, or links drawn by researchers that aren't backed up by their studies.

Be aware that the media wants to make you read the article (they get paid to do that), so they will have strong headlines, or will make unconfirmed "cause-effect" statements because it is their job to sell the story. It is *your* job to be wary.

Generalizing results to people beyond the scope of the study

You can make conclusions only about the population that's represented by your sample. If you sample men only, you can't make conclusions about women. If you sample healthy young people, you can't make your conclusions about everyone. But many researchers try to do just that, and it can give misleading results.

Here's how you can determine whether a researcher's conclusions measure up (Chapter 16 has more on samples and populations):

1. **Find out what the target population is (that is, the group that the researcher wants to make conclusions about).**

2. **Find out how the sample was selected and see whether the sample is representative of that target population (and not some more narrowly defined population).**

3. **Check the conclusions made by the researchers and make sure they're not trying to apply their results to a broader population than they actually studied.**

Making Informed Decisions

Just because someone says they conducted a "scientific study" or a "scientific experiment" doesn't mean it was done right or that the results are credible (not that I'm saying you should discount everything that you see and hear). Unfortunately, I've come across a lot of bad experiments in my days as a statistical consultant. The worst part is that if an experiment was done poorly, you can't do anything about it after the fact except ignore the results — and that's exactly what you need to do.

Here are some tips that help you make an informed decision about whether to believe the results of an experiment, especially one whose results are very important to you:

- ✔ **When you first hear or see the result, grab a pencil and write down as much as you can about what you heard or read, where you heard or read it, who did the research, and what the main results were.** (I keep pencil and paper in my TV room and in my purse just for this purpose.)

- ✔ **Follow up on your sources until you find the person who did the original research and then ask them for a copy of the report or paper.**

✔ **Go through the report and evaluate the experiment according to the eight steps for a good experiment described in the "Designing a Good Experiment" section of this chapter.** (You really don't have to understand everything written in a report in order to do that.)

✔ **Carefully scrutinize the conclusions that the researcher makes regarding his or her findings.** Many researchers tend to overstate results, make conclusions beyond the statistical evidence, or try to apply their results to a broader population than the one they studied.

✔ **Never be afraid to ask questions of the media, the researchers, and even your own experts.** For example, if you have a question about a medical study, ask your doctor. He or she will be glad that you're an empowered and well-informed patient!

✔ **And finally, don't get overly skeptical, just because you're now a lot more aware of all the bad practices going on out there.** Not everything is bad. There are many more good researchers, credible results, and well-informed reporters than not. You have to maintain a sense of being cautious and ready to spot problems without discounting everything.

Chapter 18

Looking for Links: Correlation and Regression

. . . .

. . . .

*T*oday's media provide a steady stream of information, including reports on all the latest links that have been found by researchers. Just today I heard that increased video game use can negatively affect a child's attention span, the amount of a certain hormone in a woman's body can predict when she will enter menopause, and the more depressed you get, the more chocolate you eat, and the more chocolate you eat, the more depressed you get (how depressing!).

Some studies are truly legitimate and help improve the quality and longevity of our lives. Other studies are not so clear. For example, one study says that exercising 20 minutes three times a week is better than exercising 60 minutes one time a week, another study says the opposite, and yet another study says there is no difference.

If you are a confused consumer when it comes to links and correlations, take heart; this chapter can help. You'll gain the skills to dissect and evaluate research claims and make your own decisions about those headlines and sound bites that you hear each day alerting you to the latest correlation. You'll discover what it truly means for two variables to be correlated, when a cause-and-effect relationship can be concluded, and when and how to predict one variable based on another.

Picturing a Relationship with a Scatterplot

An article in *Garden Gate* magazine caught my eye: "Count Cricket Chirps to Gauge Temperature." According to the article, all you have to do is find a cricket, count the number of times it chirps in 15 seconds, add 40, and voilà! You've just estimated the temperature in Fahrenheit.

The National Weather Service Forecast Office even puts out its own "Cricket Chirp Converter." You enter the number of cricket chirps recorded in 15 seconds, and the converter gives you the estimated temperature in four different units, including Fahrenheit and Celsius.

A fair amount of research does support the claim that frequency of cricket chirps is related to temperature. For the purpose of illustration I've taken only a subset of some of the data (see Table 18-1).

Table 18-1	Cricket Chirps and Temperature Data (Excerpt)
Number of Chirps (in 15 Seconds)	*Temperature (Fahrenheit)*
18	57
20	60
21	64
23	65
27	68
30	71
34	74
39	77

Notice that each observation is composed of two variables that are tied together: the number of times the cricket chirped in 15 seconds (the *X*-variable) and the temperature at the time the data was collected (the *Y*-variable). Statisticians call this type of two-dimensional data *bivariate* data. Each observation contains one pair of data collected simultaneously. For example, row one of Table 18-1 depicts a pair of data (18, 57).

Bivariate data is typically organized in a graph that statisticians call a *scatterplot*. A scatterplot has two dimensions, a horizontal dimension (the *X*-axis) and a vertical dimension (the *Y*-axis). Both axes are numerical; each one contains a number line. In the following sections, I explain how to make and interpret a scatterplot.

Making a scatterplot

Placing observations (or points) on a scatterplot is similar to playing the game Battleship. Each observation has two coordinates; the first corresponds to the first piece of data in the pair (that's the *X* coordinate; the amount that you go left or right). The second coordinate corresponds to the second piece of data in the pair (that's the *Y*-coordinate; the amount that you go up or down). You place the point representing that observation at the intersection of the two coordinates.

Figure 18-1 shows a scatterplot for the cricket chirps and temperature data listed in Table 18-1. Because I ordered the data according to their *X*-values, the points on the scatterplot correspond from left to right to the observations given in Table 18-1, in the order listed.

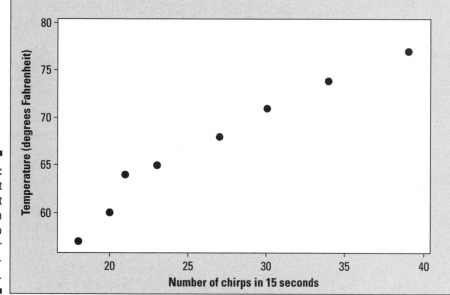

Figure 18-1: Scatterplot of cricket chirps in relation to outdoor temperature.

Interpreting a scatterplot

You interpret a scatterplot by looking for trends in the data as you go from left to right:

- ✔ If the data show an uphill pattern as you move from left to right, this indicates a *positive relationship between X and Y*. As the *X*-values increase (move right), the *Y*-values increase (move up) a certain amount.

✔ If the data show a downhill pattern as you move from left to right, this indicates a *negative relationship between X and Y.* As the *X*-values increase (move right) the *Y*-values decrease (move down) by a certain amount.

✔ If the data don't seem to resemble any kind of pattern (even a vague one), then no relationship exists between *X* and *Y*.

One pattern of special interest is a *linear* pattern, where the data has a general look of a line going uphill or downhill. Looking at Figure 18-1, you can see that a positive linear relationship does appear between number of cricket chirps and the temperature. That is, as the cricket chirps increase, the temperature increases as well.

In this chapter I explore linear relationships only. A *linear relationship between X and Y* exists when the pattern of *X*- and *Y*-values resembles a line, either uphill (with a positive slope) or downhill (with a negative slope). Other types of trends may exist in addition to the uphill/downhill linear trends (for example, curves or exponential functions); however, these trends are beyond the scope of this book. The good news is that many relationships do fall under the uphill/downhill linear scenario.

Scatterplots show possible associations or relationships between two variables. However, just because your graph or chart shows something is going on, it doesn't mean that a cause-and-effect relationship exists.

For example, a doctor observes that people who take vitamin C each day seem to have fewer colds. Does this mean vitamin C prevents colds? Not necessarily. It could be that people who are more health conscious take vitamin C each day, but they also eat healthier, are not overweight, exercise every day, and wash their hands more often. If this doctor really wants to know if it's the vitamin C that's doing it, she needs a well-designed experiment that rules out these other factors. (See the later section "Explaining the Relationship: Correlation versus Cause and Effect" for more information.)

Quantifying Linear Relationships Using the Correlation

After the bivariate data have been organized graphically with a scatterplot (see the preceding section), and you see some type of linear pattern, the next step is to do some statistics that can quantify or measure the extent and nature of the relationship. In the following sections, I discuss *correlation,* a statistic measuring the strength and direction of a linear relationship between two variables; in particular, how to calculate and interpret correlation and understand its most important properties.

Calculating the correlation

In the earlier section "Interpreting a scatterplot," I say data that resembles an uphill line has a positive linear relationship and data that resembles a downhill line has a negative linear relationship. However, I didn't address the issue of whether or not the linear relationship was strong or weak. The strength of a linear relationship depends on how closely the data resembles a line, and of course varying levels of "closeness to a line" exist.

Can one statistic measure both the strength and direction of a linear relationship between two variables? Sure! Statisticians use the *correlation coefficient* to measure the strength and direction of the linear relationship between two numerical variables X and Y. The correlation coefficient for a sample of data is denoted by r.

Although the street definition of *correlation* applies to any two items that are related (such as gender and political affiliation), statisticians use this term only in the context of two numerical variables. The formal term for correlation is the *correlation coefficient.* Many different correlation measures have been created; the one used in this case is called the *Pearson correlation coefficient* (but from now on I'll just call it the correlation).

The formula for the correlation (r) is

$$r = \frac{1}{n-1}\left(\frac{\displaystyle\sum_{x}\sum_{y}(x-\bar{x})(y-\bar{y})}{s_x s_y}\right)$$

where n is the number of pairs of data; $\bar{x}$ and $\bar{y}$ are the sample means of all the x-values and all the y-values, respectively; and s_x and s_y are the sample standard deviations of all the x- and y-values, respectively.

Use the following steps to calculate the correlation, r, from a data set:

1. **Find the mean of all the x-values ($\bar{x}$) and the mean of all the y-values ($\bar{y}$).**

   See Chapter 5 for more on calculating the mean.

2. **Find the standard deviation of all the x-values (call it s_x) and the standard deviation of all the y-values (call it s_y).**

   See Chapter 5 to find out how to calculate the standard deviation.

3. **For each (x, y) pair in the data set, take x minus $\bar{x}$ and y minus $\bar{y}$, and multiply them together to get $(x-\bar{x})(y-\bar{y})$.**

4. **Add up all the results from Step 3.**

5. **Divide the sum by** $s_x * s_y$.

6. **Divide the result by** $n - 1$, **where** n **is the number of** (x, y) **pairs. (It's the same as multiplying by 1 over** $n - 1$.)

   This gives you the correlation, r.

For example, suppose you have the data set (3, 2), (3, 3), and (6, 4). You calculate the correlation coefficient r via the following steps. (Note for this data the x-values are 3, 3, 6, and the y-values are 2, 3, 4.)

1. $\bar{x}$ **is** $12 \div 3 = 4$, **and** $\bar{y}$ **is** $9 \div 3 = 3$.

2. **The standard deviations are** $s_x = 1.73$ **and** $s_y = 1.00$.

   See Chapter 5 for step-by-step calculations.

3. **The differences found in Step 3 multiplied together are:** $(3 - 4)(2 - 3) = (-1)(-1) = +1; (3 - 4)(3 - 3) = (-1)(0) = 0; (6 - 4)(4 - 3) = (2)(1) = +2$.

4. **Adding the Step 3 results, you get** $1 + 0 + 2 = 3$.

5. **Dividing by** $s_x * s_y$ **gives you** $3 \div (1.73 * 1.00) = 3 \div 1.73 = 1.73$.

6. **Now divide the Step 5 result by** $3 - 1$ **(which is 2), and you get the correlation** $r = 0.87$.

Interpreting the correlation

The correlation r is always between +1 and –1. To interpret various values of r (no hard and fast rules here, just Rumsey's rule of thumb), see which of the following values your correlation is closest to:

- ✔ **Exactly –1:** A perfect downhill (negative) linear relationship

- ✔ **–0.70:** A strong downhill (negative) linear relationship

- ✔ **–0.50:** A moderate downhill (negative) relationship

- ✔ **–0.30:** A weak downhill (negative) linear relationship

- ✔ **0:** No linear relationship

- ✔ **+0.30:** A weak uphill (positive) linear relationship

- ✔ **+0.50:** A moderate uphill (positive) relationship

- ✔ **+0.70:** A strong uphill (positive) linear relationship

- ✔ **Exactly +1:** A perfect uphill (positive) linear relationship

If the scatterplot doesn't indicate there's at least somewhat of a linear relationship, the correlation doesn't mean much. Why measure the amount of linear relationship if there isn't enough of one to speak of? However you can take the idea of no linear relationship two ways: 1) If no relationship at all exists, calculating the correlation doesn't make sense because correlation only applies to linear relationships; and 2) If a strong relationship exists but it's not linear, the correlation may be misleading, because in some cases a strong curved relationship exists yet the correlation turns out to be strong. That's why it's critical to examine the scatterplot first.

Figure 18-2 shows examples of what various correlations look like, in terms of the strength and direction of the relationship. Figure 18-2a shows a correlation of +1, Figure 18-2b shows a correlation of –0.50, Figure 18-2c shows a correlation of +0.85, and Figure 18-2d shows a correlation of +0.15. Comparing Figures 18-2a and c, you see Figure 18-2a is a perfect uphill straight line, and Figure 18-2c shows a very strong uphill linear pattern. Figure 18-2b is going downhill but the points are somewhat scattered in a wider band, showing a linear relationship is present, but not as strong as in Figures 18-2a and 18-2c. Figure 18-2d doesn't show much of anything happening (and it shouldn't, since its correlation is very close to 0).

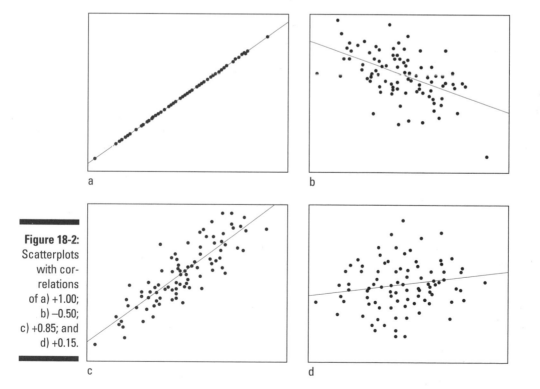

Figure 18-2:
Scatterplots with correlations of a) +1.00; b) –0.50; c) +0.85; and d) +0.15.

a

b

c

d

Many folks make the mistake of thinking that a correlation of –1 is a bad thing, indicating no relationship. Just the opposite is true! A correlation of –1 means the data are lined up in a perfect straight line, the strongest linear relationship you can get. The "–" (minus) sign just happens to indicate a negative relationship, a downhill line.

How close is close enough to –1 or +1 to indicate a strong enough linear relationship? Most statisticians like to see correlations beyond at least +0.5 or –0.5 before getting too excited about them. Don't expect a correlation to always be 0.99 however; remember, this is real data, and real data aren't perfect.

For my subset of the cricket chirps versus temperature data from the earlier section "Picturing a Relationship with a Scatterplot," I calculated a correlation of 0.98, which is almost unheard of in the real world (these crickets are *good!*).

Examining properties of the correlation

Here are several important properties of the correlation coefficient:

- ✔ The correlation is always between –1 and +1, as I explain in the preceding section.

- ✔ The correlation is a unitless measure, which means that if you change the units of X or Y, the correlation won't change. For example, changing the temperature from Fahrenheit to Celsius won't affect the correlation between the frequency of chirps (X) and the outside temperature (Y).

- ✔ The variables X and Y can be switched in the data set without changing the correlation. For example, if height and weight have a correlation of 0.53, weight and height have the same correlation.

Working with Linear Regression

In the case of two numerical variables X and Y, when at least a moderate correlation has been established through both the correlation and the scatterplot, you know they have some type of linear relationship. Researchers often use that relationship to predict the (average) value of Y for a given value of X using a straight line. Statisticians call this line the *regression line*. If you know the slope and the y-intercept of that regression line, then you can plug in a value for X and predict the average value for Y. In other words, you predict (the average) Y from X. In the following sections, I provide the basics of understanding and using the linear regression equation (I explain how to make predictions with linear regression later in this chapter).

Never do a regression analysis unless you have already found at least a moderately strong correlation between the two variables. (My rule of thumb is it should be at or beyond either positive or negative 0.50, but other statisticians may have different criteria.) I've seen cases where researchers go ahead and make predictions when a correlation is as low as 0.20! By anyone's standards, that doesn't make sense. If the data don't resemble a line to begin with, you shouldn't try to use a line to fit the data and make predictions (but people still try).

Figuring out which variable is X and which is Y

Before moving forward to find the equation for your regression line, you have to identify which of your two variables is X and which is Y. When doing correlations (as I explain earlier in this chapter), the choice of which variable is X and which is Y doesn't matter, as long as you're consistent for all the data. But when fitting lines and making predictions, the choice of X and Y does make a difference.

So how do you determine which variable is which? In general, Y is the variable that you want to predict, and X is the variable you are using to make that prediction. In the earlier cricket chirps example, you are using the number of chirps to predict the temperature. So in this case the variable Y is the temperature, and the variable X is the number of chirps. Hence Y can be predicted by X using the equation of a line if a strong enough linear relationship exists.

Statisticians call the X-variable (cricket chirps in my earlier example) the *explanatory variable,* because if X changes, the slope tells you (or explains) how much Y is expected to change in response. Therefore, the Y variable is called the *response variable.* Other names for X and Y include the *independent* and *dependent* variables, respectively.

Checking the conditions

In the case of two numerical variables, you can come up with a line that enables you to predict Y from X, if (and only if) the following two conditions from the previous sections are met:

- ✔ The scatterplot must form a linear pattern.
- ✔ The correlation, r, is moderate to strong (typically beyond 0.50 or –0.50).

Some researchers actually don't check these conditions before making predictions. Their claims are not valid unless the two conditions are met.

But suppose the correlation is high; do you still need to look at the scatterplot? Yes. In some situations the data have a somewhat curved shape, yet the correlation is still strong; in these cases making predictions using a straight line is still invalid. Predictions need to be made based on a curve. (This topic is outside the scope of this book; if you are interested, see *Statistics II For Dummies,* where I tackle nonlinear relationships.)

Calculating the regression line

For the crickets and temperature data, you can see that the scatterplot in Figure 18-1 shows a linear pattern. The correlation between cricket chirps and temperature was found earlier in this chapter to be very strong ($r = 0.98$). You now can find one line that best fits the data (in terms of having the smallest overall distance to the points). Statisticians call this technique for finding the best-fitting line a *simple linear regression analysis using the least squares method.*

The formula for the *best-fitting line* (or *regression line*) is $y = mx + b$, where m is the slope of the line and b is the y-intercept. This equation itself is the same one used to find a line in algebra; but remember, in statistics the points don't lie perfectly on a line — the line is a model around which the data lie if a strong linear pattern exists.

- The *slope* of a line is the change in Y over the change in X. For example, a slope of $^{10}\!/_3$ means as the x-value increases (moves right) by 3 units, the y-value moves up by 10 units on average.

- The *y-intercept* is that place on the y-axis where the value of x is zero. For example, in the equation $2x - 6$, the line crosses the y-axis at the point -6. The coordinates of this point are $(0, -6)$; when a line crosses the y-axis, the x-value is always 0.

To come up with the best-fitting line, you need to find values for m and b that fit the pattern of data the best, for your given criteria. Different criteria exist and can lead to other lines, but the criteria I use in this book (and in all introductory level statistics courses in general) is to find the line that minimizes what statisticians call the *sum of squares for error (SSE)*. The SSE is the sum of all the squared differences from the points on the proposed line to the actual points in the data set. The line with the lowest possible SSE wins and its equation is used as the best-fitting line. This process is where the name *the least-squares method* comes from.

You may be thinking that you have to try lots and lots of different lines to see which one fits best. Fortunately, you have a more straightforward option (although eyeballing a line on the scatterplot does help you think about what you'd expect the answer to be). The best-fitting line has a distinct slope and *y*-intercept that can be calculated using formulas (and, I may add, these formulas aren't too hard to calculate).

To save a great deal of time calculating the best fitting line, first find the "big five," five summary statistics that you'll need in your calculations:

1. The mean of the x values (denoted $\bar{x}$)

2. The mean of the y values (denoted $\bar{y}$)

3. The standard deviation of the x values (denoted s_x)

4. The standard deviation of the y values (denoted s_y)

5. The correlation between X and Y (denoted r)

Finding the slope

The formula for the slope, m, of the best-fitting line is

$$m = r\left(\frac{s_y}{s_x}\right)$$

where r is the correlation between X and Y, and s_x and s_y are the standard deviations of the x-values and the y-values, respectively. You simply divide s_y by s_x and multiply the result by r.

Note that the slope of the best-fitting line can be a negative number because the correlation can be a negative number. A negative slope indicates that the line is going downhill. For example, an increase in police officers is related to a decrease in the number of crimes in a linear fashion; the correlation and hence the slope of the best-fitting line is negative in this case.

The correlation and the slope of the best-fitting line are not the same. The formula for slope takes the correlation (a unitless measurement) and attaches units to it. Think of $s_y \div s_x$ as the variation (resembling change) in Y over the variation in X, in units of X and Y. For example, variation in temperature (degrees Fahrenheit) over the variation in number of cricket chirps (in 15 seconds).

Finding the y-intercept

The formula for the y-intercept, b, of the best-fitting line is $b = \bar{y} - m\bar{x}$, where $\bar{x}$ and $\bar{y}$ are the means of the x-values and the y-values, respectively, and m is the slope (the formula for which is given in the preceding section).

So to calculate the y-intercept, b, of the best-fitting line, you start by finding the slope, m, of the best-fitting line using the steps listed in the preceding section. You then multiply m by $\bar{x}$ and subtract your result from $\bar{y}$.

Always calculate the slope before the *y*-intercept. The formula for the *y*-intercept contains the slope!

Interpreting the regression line

Even more important than being able to calculate the slope and *y*-intercept to form the best-fitting regression line is the ability to interpret their values; I explain how to do so in the following sections.

Interpreting the slope

The slope is interpreted in algebra as *rise over run*. If, for example, the slope is 2, you can write this as ²⁄₁ and say that as you move from point to point on the line, as the value of the *X* variable increases by 1, the value of the *Y* variable increases by 2. In a regression context, the slope is the heart and soul of the equation because it tells you how much you can expect *Y* to change as *X* increases.

In general, the units for slope are the units of the *Y* variable per units of the *X* variable. It's a ratio of change in *Y* per change in *X*. Suppose in studying the effect of dosage level in milligrams (mg) on systolic blood pressure (mmHg), a researcher finds that the slope of the regression line is –2.5. You can write this as ⁻²·⁵⁄₁ and say that systolic blood pressure is expected to decrease by 2.5 mmHg on average per 1 mg increase in drug dosage.

Always make sure to use proper units when interpreting slope. If you don't consider units, you won't really see the connection between the two variables at hand. For example if *Y* is exam score and *X* = study time, and you find the slope of the equation is 5, what does this mean? Not much without any units to draw from. Including the units, you see you get an increase of 5 points (change in *Y*) for every 1 hour increase in studying (change in *X*). Also be sure to watch for variables that have more than one common unit, such as temperature being in either Fahrenheit or Celsius; know which unit is being used.

If using a 1 in the denominator of slope is not super-meaningful to you, you can multiply the top and bottom by any number (as long as it's the same number) and interpret it that way instead. In the systolic blood pressure example, instead of writing slope as ⁻²·⁵⁄₁ and interpreting it as a drop of 2.5 mmHg per 1 mg increase of the drug, you can multiply the top and bottom by 10 to get ⁻²⁵⁄₁₀ and say an increase in dosage of 10 mg results in a decrease in systolic blood pressure of 25 mmHg.

Interpreting the y-intercept

The *y*-intercept is the place where the regression line $y = mx + b$ crosses the *y*-axis where $x = 0$, and is denoted by *b* (see the earlier section "Finding the *y*-intercept"). Sometimes the *y*-intercept can be interpreted in a meaningful way, and sometimes not. This uncertainty differs from slope, which is always interpretable. In fact, between the two elements of slope and *y*-intercept, the slope is the star of the show, with the *y*-intercept serving as the less-famous but still noticeable sidekick.

At times the *y*-intercept makes no sense. For example, suppose you use rain to predict bushels per acre of corn. You know if the data set contains a point where rain is 0, the bushels per acre must be 0 as well. As a result, if the regression line crosses the *y*-axis somewhere else besides 0 (and there is no guarantee it will cross at 0 — it depends on the data), the *y*-intercept will make no sense. Similarly, in this context a negative value of *y* (corn production) cannot be interpreted.

Another situation where you can't interpret the *y*-intercept is when data are not present near the point where $x = 0$. For example, suppose you want to use students' scores on Midterm 1 to predict their scores on Midterm 2. The *y*-intercept represents a prediction for Midterm 2 when the score on Midterm 1 is 0. You don't expect scores on a midterm to be at or near 0 unless someone didn't take the exam, in which case her score wouldn't be included in the first place.

Many times, however, the *y*-intercept is of interest to you, it has meaning, and you have data collected in the area where $x = 0$. For example, if you're predicting coffee sales at football games in Green Bay, Wisconsin, using temperature, some games get cold enough to have temperatures at or even below 0 degrees Fahrenheit, so predicting coffee sales at these temperatures makes sense. (As you may guess, they sell more and more coffee as the temperature dips.)

Putting it all together with an example: The regression line for the crickets

In the earlier section "Picturing a Relationship with a Scatterplot," I introduce the example of cricket chirps related to temperature. The "big five" statistics, which I explain in "Calculating the regression line," are shown in Table 18-2 for the subset of cricket data. (*Note:* I'm rounding for ease of explanation only.)

Table 18-2	"Big Five" Statistics for the Cricket Data		
Variable	*Mean*	*Standard Deviation*	*Correlation*
Number of chirps (*x*)	$\bar{x} = 26.5$	$s_x = 7.4$	$r = +0.98$
Temp (*y*)	$\bar{y} = 67$	$s_y = 6.8$	

The slope, m, for the best-fitting line for the subset of cricket chirps versus temperature data is $m = r\dfrac{s_y}{s_x} = 0.98\left(\dfrac{6.8}{7.4}\right) = 0.90$. So as the number of chirps increases by 1 chirp per 15 seconds, the temperature is expected to increase by 0.90 degrees Fahrenheit on average. To get a more meaningful interpretation, you can multiply the top and bottom of the slope by 10 and say as chirps increase by 10 (per 15 seconds) temperature increases 9 degrees Fahrenheit.

Now, to find the y-intercept, b, you take $\bar{y} - m\bar{x}$, or $67 - (0.90)(26.5) = 43.15$. So the best-fitting line for predicting temperature from cricket chirps based on the data is $y = 0.90x + 43.15$, or temperature (in degrees Fahrenheit) = 0.90 * (number of chirps in 15 seconds) + 43.2. Now can you use the y-intercept to predict temperature when no chirping is going on at all? Because no data was collected at or near this point, you cannot make predictions for temperature in this area. You can't predict temperature using crickets if the crickets are silent.

Making Proper Predictions

After you have determined a strong linear relationship and you find the equation of the best fitting line using $y = mx + b$, you use that line to predict (the average) y for a given x-value. To make predictions, you plug the x-value into the equation and solve for y. For example, if your equation is $y = 2x + 1$ and you want to predict y for $x = 1$, then plug 1 into the equation for x to get $y = 2(1) + 1 = 3$.

Keep in mind that you choose the values of X (the explanatory variable) that you plug in; what you predict is Y, the response variable, which totally depends on X. By doing this, you are using one variable that you can easily collect data on to predict a Y variable that is difficult or not possible to measure. This process works well as long as X and Y are correlated. This concept is the big idea of regression.

Using the examples from the previous section, the best-fitting line for the crickets is $y = 0.90x + 43.2$. Say you're camping outside, listening to the crickets, and remember you can predict temperature by counting cricket chirps. You count 35 chirps in 15 seconds, put in 35 for x, and find that $y = 0.9(35) + 43.2 = 74.7$. (Yeah, you memorized the formula before you went camping just in case you needed it.) So because the crickets chirped 35 times in 15 seconds, you figure the temperature is probably about 75 degrees Fahrenheit.

Just because you have a regression line doesn't mean you can plug in *any* value for X and do a good job of predicting Y. Making predictions using x-values that fall outside the range of your data is a no-no. Statisticians call this *extrapolation;* watch for researchers who try to make claims beyond the range of their data.

For example, in the chirping data, no data is collected for fewer than 18 chirps or more than 39 chirps per 15 seconds (refer to Table 18-1). If you try to make predictions outside this range, you are going into uncharted territory; the farther outside this range you go with your *x*-values, the more dubious your predictions for *y* will get. Who's to say the line still works outside of the area where data were collected? Do you really think that crickets will chirp faster and faster without limit? At some point they would either pass out or burn up! And what does a negative number of chirps really mean? (Is this similar to asking what the sound of one hand clapping is?)

Be aware that not every data point will necessarily fit the regression line well, even if the correlation is high. A point or two may fall outside the overall pattern of the rest of the data; such points are called *outliers*. One or two outliers probably won't affect the overall fit of the regression line much, but in the end you can see that the line didn't do well at those specific points.

The numerical difference between the predicted value of *y* from the line and the actual *y*-value you got from your data is called a *residual*. Outliers have large residuals compared to the rest of the points; they are worth investigating to see if there was an error in the data at those points or if there is something particularly interesting in the data to follow up on. (I give a much more detailed look at residuals in the book *Statistics II For Dummies*.)

Explaining the Relationship: Correlation versus Cause and Effect

Scatterplots and correlations identify and quantify relationships between two variables. However, if a scatterplot shows a definite pattern and the data are found to have a strong correlation, that doesn't necessarily mean that a cause-and-effect relationship exists between the two variables. A *cause-and-effect relationship* is one where a change in one variable (in this case *X*) causes a change in another variable (in this case *Y*). (In other words, the change in *Y* is not only associated with a change in *X*, but also directly caused by *X*.)

For example, suppose a well-controlled medical experiment is conducted to determine the effects of dosage of a certain drug on blood pressure. (See a total breakdown of experiments in Chapter 17.) The researchers look at their scatterplot and see a definite downhill linear pattern; they calculate the correlation, and it's strong. They conclude that increasing the dosage of this drug causes a decrease in blood pressure. This cause-and-effect conclusion is okay because they controlled for other variables that could affect blood pressure in their experiment, such as other drugs taken, age, general health, and so on.

However, if you made a scatterplot and examined the correlation between ice cream consumption versus murder rates in New York City, you would also see a strong linear relationship (this one is uphill). Yet no one would claim that more ice cream consumption causes more murders to occur.

What's going on here? In the first case, the data were collected through a well-controlled medical experiment, which minimizes the influence of other factors that may affect blood pressure. In the second example, the data were based just on observation, and no other factors were examined. Researchers subsequently found out that this strong relationship exists because increases in murder rates and ice cream sales are both related to increases in temperature. Temperature in this case is called a *confounding variable*; it affects both *X* and *Y* but was not included in the study (see Chapter 17).

Whether two variables are found to be causally associated depends on how the study was conducted. I've seen many instances in which people try to claim cause-and-effect relationships just by looking at scatterplots or correlations. Why would they do this? Because they want to believe it (in other words for them it's "believing is seeing," rather than the other way around). Beware of this tactic. In order to establish cause and effect, you need to have a well-designed experiment or a boatload of observational studies. If someone is trying to establish a cause-and-effect relationship by showing a chart or graph, dig deeper to find out how the study was designed and how the data were collected, and evaluate the study appropriately using the criteria outlined in Chapter 17.

The need for a well-designed experiment in order to claim cause and effect is often ignored by some researchers and members of the media, who give us headlines such as "Doctors can lower malpractice lawsuits by spending more time with patients." In reality, it was found that doctors who have fewer lawsuits are the type who spend a lot of time with patients. But that doesn't mean taking a bad doctor and having him spend more time with his patients will reduce his malpractice suits; in fact, spending more time with them may create even more problems.

Chapter 19

Two-Way Tables and Independence

*C*ategorical variables place individuals into groups based on certain characteristics, behaviors, or outcomes, such as whether you ate breakfast this morning (yes, no) or political affiliation (Democrat, Republican, Independent, "other"). Oftentimes people look for relationships between two categorical variables; hardly a day goes by that you don't hear about another relationship that's reported to have been found.

Here are just a few examples I found on the Internet recently:

✔ Dog owners are more likely to take their animal to the vet than cat owners.

✔ Heavy use of social-networking Web sites in teens is linked to depression.

✔ Children who play more video games do better in science classes.

With all this information being given to you about variables that are related, how do you decide what to believe? For example, does heavy use of social-networking Web sites cause depression, or is it the other way around? Or perhaps a third variable out there is related to both of them, such as problems in the home.

In this chapter, you see how to organize and analyze data from two categorical variables. You find out how to use proportions to make comparisons and look at overall patterns and how to check for independence of two categorical variables. You see how to describe dependent relationships appropriately and to evaluate results claiming to indicate cause-and-effect relationships, making predictions, and/or projecting their results to a population.

Organizing a Two-Way Table

To explore links between two categorical variables, you first need to organize the data that's been collected, and a table is a great way to do that. A *two-way table* classifies individuals into groups based on the outcomes of two categorical variables (for example, gender and opinion).

Suppose your local community developers are building a campground, and they've decided pets will be allowed as long as they're on a leash. They are now trying to decide whether the campground should have a separate section for pets. You have a hunch that non–pet campers in the area may be more in favor of a separate pet area than pet campers, so you decide to find out what the members of the camping community think. You randomly select 100 campers from the local area and conduct a pet camping survey, recording each person's opinion on having a pet section (yes, no) and if they camp with pets (yes, no). You now have a spreadsheet with 100 rows of data, one for each person you surveyed. Each row has two pieces of data: one column for whether the person is a pet camper (yes, no) and one column for that person's opinion on having a pet section (support, oppose). Suppose the first 10 rows of your data set look like what's shown in Table 19-1.

Table 19-1	First 10 Rows of Data from the Pet Camping Survey	
Person	*Pet Camper?*	*Opinion on a Separate Pet Section*
1	Yes	Oppose
2	Yes	Oppose
3	Yes	Support
4	No	Support
5	No	Support
6	Yes	Support
7	No	Oppose
8	No	Support
9	Yes	Support
10	No	Oppose

From this small portion of your data set, you can start to break it down yourself. For example, looking at column 2 results, you see that half the respondents (5 ÷ 10 = 0.50) camp with pets and the other half do not. Of those who camp with pets (that is, of those five people who have a yes in column 2), three of them (60%) support having a separate section; and the same results are true for non–pet campers. These results from these 10 campers likely don't apply to all 100 campers surveyed; however, if you tried to examine the raw data from all 100 rows of this data set by hand, you wouldn't make much progress in seeing patterns without a lot of hard work.

In order to get a handle on what's happening in a large data set when you are examining two categorical variables, you organize your data into a two-way table. The following sections take you through it.

Setting up the cells

A two-way table organizes categorical data from two variables by using rows to represent one variable (such as pet camping — yes or no) and columns to represent the other variable (such as opinion on a pet section — support or oppose). Each person appears exactly once in the table.

Continuing with the camping example I start earlier in this chapter, in Table 19-2 I summarize the results from all 100 campers surveyed.

Table 19-2	Two-Way Table of Pet Camping Survey Data (All 100 Rows)	
	Support Separate Pet Section	*Oppose Separate Pet Section*
Pet Camper	20	10
Non–Pet Camper	55	15

Table 19-2 has 2 $*$ 2 = 4 numbers in it. These numbers represent the *cells* of the two-way table; each one represents an intersection of a row and column. The cell in the upper left corner of the table represents the 20 people who are pet campers supporting a pet section. In the upper right cell 10 people are pet campers opposing a pet section. In the lower left are the 55 non–pet campers who want a pet section; the 15 people in the lower right are non–pet campers opposing a pet section.

Figuring the totals

Before getting to the nitty-gritty analysis of a two-way table in the later section "Interpreting Results from a Two-Way Table," you calculate some totals and add them to the table for later reference. You summarize each variable separately by calculating the *marginal totals*, which represent the total number in each row (for the first variable) and the total number in each column (for the second variable). The *marginal row totals* form an additional column on the right side of the table, and the *marginal column totals* form an additional row on the bottom of the table.

For example, in Table 19-2 in the preceding section, the marginal row total for row 1, the number of pet campers, is 20 + 10 = 30; the marginal row total for non–pet campers (row 2) is 55 + 15 = 70. The marginal column total for those wanting a pet section (column 1) is 20 + 55 = 75; and the marginal column total for those not wanting a separate section (column 2) is 10 + 15 = 25.

The *grand total* is the total of all the cells in the table and is equal to the sample size. (Note the marginal totals are not included in the grand total, only the cells.) The grand total sits in the lower right-hand corner of the two-way table. In this example, the grand total is 20 + 10 + 55 + 15 = 100. Table 19-3 shows the marginal row and column totals and the grand total for the pet camping survey data.

The marginal row totals always sum to the grand total, because everyone in the survey either camps with a pet or they don't. In the last column of Table 19-3 you see that 30 + 70 = 100. Similarly the marginal column totals always sum to the grand total; everyone in the survey either wants a pet section or they don't; in the last row of Table 19-3 you see 75 + 25 = 100.

Table 19-3	Two-Way Table of Pet Camping Survey Data, Including Marginal Totals		
	Support Separate Pet Section	**Oppose Separate Pet Section**	**Marginal Row Totals**
Pet Camper	20	10	20 + 10 = 30
Non–Pet Camper	55	15	55 + 15 = 70
Marginal Column Totals	20 + 55 = 75	10 + 15 = 25	**Grand total = 100** (20 + 10 + 55 + 15)

When organizing a two-way table, always include the marginal totals and the grand total. It gets you off on the right foot when analyzing the data.

Interpreting Two-Way Tables

After the two-way table is set up (with the help of the information in the previous section), you calculate percents to explore the data to answer your research questions. Here are some questions of interest from the camping data earlier in this chapter (each question will be handled in the following sections, respectively):

- ✔ What percentage of the campers are in favor of a pet section?
- ✔ What percentage of the campers are pet campers who support a pet section?
- ✔ Do more non–pet campers support a pet section, compared to pet campers?

The answers to these (and any other) questions about the data come from finding and working with the proportions, or percentages, of individuals within certain parts of the table. This process involves calculating and examining what statisticians call *distributions*. A distribution in the case of a two-way table is a list of all the possible outcomes for one variable or a combination of variables, along with their corresponding proportions (or percentages).

For example, the distribution for the pet camping variable lists the percentages of people who do and do not camp with pets. The distribution for the combination of the pet camping variable (yes, no) and the opinion variable (support, oppose) lists the percentages of: 1) pet campers who support a pet section; 2) pet campers who oppose a pet section; 3) non–pet campers who support a pet section; and 4) the non–pet campers who oppose a pet section.

For any distribution, all the percentages must sum to 100%. If you're using proportions (decimals), they must sum to 1.00. Each individual has to be somewhere, and he can't be in more than one place at one time.

In the following sections, you see how to find three types of distributions, each one helping you to answer its corresponding question in the preceding list.

Singling out variables with marginal distributions

If you want to examine one variable at a time in a two-way table, you don't look in the cells of the table, but rather in the margins. As seen in the earlier section "Figuring the totals," the marginal totals represent the total number in each row (or column) separately. In the two-way table for the pet camping survey (refer to Table 19-3), you see the marginal totals for the pet camping variable (yes/no) in the right-hand column, and you find the marginal totals for the opinion variable (support/oppose) in the bottom row.

If you want to make comparisons between two groups (for example, pet campers versus non–pet campers), however, the results are easier to interpret if you use proportions instead of totals. If 350 people were surveyed, visualizing a comparison is easier if you're told that 60% are in Group A and 40% are in Group B, rather than saying 210 people are in Group A and 140 are in Group B.

To examine the results of a two-way table based on a single variable, you find what statisticians call the *marginal distribution* for that variable. In the following sections, I show you how to calculate and graph marginal distributions.

Calculating marginal distributions

To find a marginal distribution for one variable in a two-way table, you take the marginal total for each row (or column) divided by the grand total.

- If your variable is represented by the rows (for example, the pet camping variable in Table 19-3), use the marginal row totals in the numerators and the grand total in the denominators. Table 19-4 shows the marginal distribution for the pet camping variable (yes, no).

- If your variable is represented by the columns (for example, opinion on the pet section policy, shown in Table 19-3), use the marginal column totals for the numerators and the grand total for the denominators. Table 19-5 shows the marginal distribution for the opinion variable (support, oppose).

In either case, the sum of the proportions for any marginal distribution must be 1 (subject to rounding). All results in a two-way table are subject to rounding error; to reduce rounding error, keep at least 2 digits after the decimal point throughout.

Table 19-4 Marginal Distribution for Pet Camping Variable

Pet Camping	Proportion
Yes	$30 \div 100 = 0.30$
No	$70 \div 100 = 0.70$
Total	1.00

Table 19-5 Marginal Distribution for the Opinion Variable

Opinion	Proportion
Support pet section	$75 \div 100 = 0.75$
Oppose pet section	$25 \div 100 = 0.25$
Total	1.00

Graphing marginal distributions

You graph a marginal distribution using either a pie chart or a bar graph. Each graph shows the proportion of individuals within each group for a single variable. Figure 19-1a is a pie chart summarizing the pet camping

variable, and Figure 19-1b is a pie chart showing the breakdown of the opinion variable. You see that the results of these two pie charts correspond with
the marginal distributions in Tables 19-4 and 19-5, respectively.

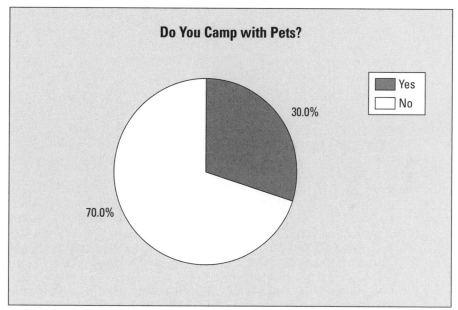

a

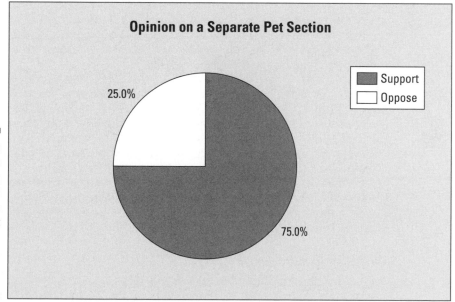

Figure 19-1:
Pie charts
showing
marginal
distributions
for a) pet
camping
variable;
and b) opin-
ion variable.

b

From the results of the two separate marginal distributions for the pet camping and opinion variables, you say that the majority of all the campers in this sample are non–pet campers (70%) and the majority of all the campers in this sample (75%) support the idea of having a pet section.

While marginal distributions show us how each variable breaks down on its own, they don't tell us about the connection between two variables. For the camping example, you know what percentage of all campers support a new pet section, but you can't distinguish the opinions of the pet campers from the non–pet campers. Distributions for making such comparisons are found in the later section, "Comparing groups with conditional distributions."

Examining all groups — a joint distribution

Story time: A certain auto manufacturer conducted a survey to see what characteristics customers prefer in their small pickup trucks. They found that the most popular color for these trucks was red and the most popular option was four-wheel drive. In response to these results, the company started making more of their small pickup trucks red with four-wheel drive.

Guess what? They struck out; people weren't buying those trucks. Turns out that the customers who bought the red trucks were more likely to be women, and women didn't use four-wheel drive as often as men did. Customers who bought the four-wheel drive trucks were more likely to be men, and they tended to prefer black ones over red ones. So the most popular outcome of the first variable (color) paired with the most popular outcome of the second variable (options on the vehicle) doesn't necessarily add up to the most popular combination of the two variables.

To figure out which combination of two categorical variables contains the highest proportion, you need to compare the cell proportions (for example, the color and vehicle options together) rather than the marginal proportions (the color and vehicle option separately). The *joint distribution* of both variables in a two-way table is a listing of all possible row and column combinations and the proportion of individuals within each group. You use it to answer questions involving two characteristics; such as "What proportion of the voters are Democrat and female?" or, "What percentage of the campers are pet campers who support a pet section?" In the following sections, I show you how to calculate and graph joint distributions.

Calculating joint distributions

A joint distribution shows the proportion of the data that lies in each cell of the two-way table. For the pet camping example, the four row-column combinations are:

✔ All campers who camp with pets and support a pet section.

✔ All campers who camp with pets and oppose a pet section.

✔ All campers who don't camp with pets and support a pet section.

✔ All campers who don't camp with pets and oppose a pet section.

The key phrase in all of the proportions mentioned in the preceding list is *all campers.* You are taking the entire group of all campers in the survey and breaking them into four separate groups. When you see the word *all,* think joint distribution. Table 19-6 shows the joint distribution for all campers in the pet camping survey.

Table 19-6   Joint Distribution for the Pet Camping Survey Data

	Support Separate Pet Section	*Oppose Separate Pet Section*
Camp with Pets	20 ÷ 100 = 0.20	10 ÷ 100 = 0.10
Don't Camp with Pets	55 ÷ 100 = 0.55	15 ÷ 100 = 0.15

To find a joint distribution for a two-way table, you take the cell count (the number of individuals in a cell) divided by the grand total, for each cell in the table. The total of all these proportions should be 1 (subject to rounding error).

To get the numbers in the cells of Table 19-6, take the cells of Table 19-3 and divide by their corresponding grand total (100, in this case). Using the results listed in Table 19-6, you report the following:

✔ 20% of all campers surveyed camp with pets and support a pet section. (See the upper left-hand cell of the table.)

✔ 10% of all campers surveyed camp with pets and oppose a pet section. (See the upper right-hand cell of the table.)

✔ 55% of all campers surveyed don't camp with pets and do support the pet section policy. (See the lower left-hand cell of the table.)

✔ 15% of all campers surveyed don't camp with pets and oppose the pet section policy. (See the lower right-hand cell of the table.)

Adding all the proportions shown in Table 19-6, you get 0.20 + 0.10 + 0.55 + 0.15 = 1.00. Every camper shows up in one and only one of the cells of the table.

Graphing joint distributions

To graph a joint distribution from a two-way table, you make a single pie chart with four slices, representing each proportion of the data that falls within a row-column combination. Groups containing more individuals get a bigger piece of the overall pie, and hence get more weight when all the votes are counted up. Figure 19-2 is a pie chart showing the joint distribution for the pet camping survey data.

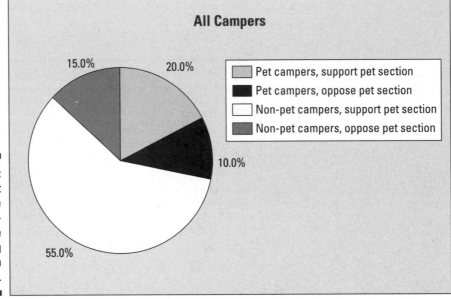

Figure 19-2:
Pie chart showing the joint distribution of the pet camping and opinion variables.

From the pie chart shown in Figure 19-2, you see some results that stand out. The majority of campers in this sample (0.55 or 55%) don't camp with pets and support a separate section for pets. The smallest slice of the pie represents those campers who camp with pets and are opposed to a separate section for pets (0.10 or 10%).

A joint distribution gives you a breakdown of the entire group by both variables at once and allows you to compare the cells to each other and to the whole group. The results in Figure 19-2 show that if they were asked to vote today as to whether or not to have a pet section, when all the votes were added up, most of the weight would be placed on the opinions of non–pet campers, because they make up the majority of campers in the survey (70%, according to Table 19-4), and the pet campers would have less of a voice, because they are a smaller group (30%).

A limitation of a joint distribution is that you can't fairly compare two groups to each other (for example pet campers versus non–pet campers) because the joint distribution puts more weight on larger groups. The next section shows how to fairly compare the groups in a two-way table.

Comparing groups with conditional distributions

You need a different type of distribution other than a joint distribution to compare the results from two groups (for example comparing opinions of pet campers versus non–pet campers). *Conditional distributions* are used when looking for relationships between two categorical variables; the individuals are first split into the groups you want to compare (for example, pet campers and non–pet campers); then the groups are compared based on their opinion on a pet section (yes, no). In the following sections, I explain how to calculate and graph conditional distributions.

Calculating conditional distributions

To find conditional distributions for the purpose of comparison, first split the individuals into groups according to the variable you want to compare. Then for each group, take the cell count (the number of individuals in a particular cell) divided by the marginal total for that group. Do this for all the cells in that group. Now repeat for the other group, using its marginal total as the denominator and the cells within its group as the numerators. (See the earlier section "Figuring the totals" for more about marginal totals.) You now have two conditional distributions, one for each group, and you fairly compare the results for the two groups.

For the pet camping survey data example (earlier in this chapter), you compare the opinions of two groups: pet campers and non–pet campers; in statistical terms you want to find the conditional distributions of opinion based on the pet camping variable. That means you split the individuals into the pet camper and non–pet camper groups, and then for each group, you find the percentages of who supports and opposes the new pet section. Table 19-7 shows these two conditional distributions in table form (working off Table 19-3).

Table 19-7	Conditional Distributions of Opinion for Pet Campers versus Non–Pet Campers		
	Support Pet Section Policy	Oppose Pet Section Policy	Total
Pet Campers	20 ÷ 30 = 0.67	10 ÷ 30 = 0.33	1.00
Non–Pet Campers	55 ÷ 70 = 0.79	15 ÷ 70 = 0.21	1.00

Notice that Table 19-7 differs from Table 19-6 in the earlier section "Calculating joint distributions" in terms of how the values in the table add up. This represents the key difference between a joint distribution and a conditional distribution that allows you to make fair comparisons using the conditional distribution:

✔ In Table 19-6, the proportions in the cells of the entire table sum to 1 because the entire group is broken down by both variables at once in a joint distribution.

✔ In Table 19-7, the proportions in each row of the table sum to 1 because each group is treated separately in a conditional distribution.

Graphing conditional distributions

One effective way to graph conditional distributions is to make a pie chart for each group (for example, one for pet campers and one for non–pet campers) where each pie chart shows the results of the variable being studied (opinion: yes or no).

Another method is to use a stacked bar graph. A *stacked bar graph* is a special bar graph where each bar has a height of 1 and represents an entire group (one bar for pet campers and one bar for non–pet campers). Each bar shows how that group breaks down regarding the other variable being studied (opinion: yes or no).

Figure 19-3 is a stacked bar graph showing two conditional distributions. The first bar is the conditional distribution of opinion for the pet camping group (row 1 of Table 19-7) and the second bar represents the conditional distribution of opinion for the non–pet camping group (row 2 of Table 19-7).

Using Table 19-7 and Figure 19-3, first look at the opinions of each group. More than 50% of the pet campers support the pet section (the exact number rounds to 67%), so you say the majority of pet campers support a pet section. Similarly, the majority of non–pet campers (about 79%, way more than half) support a pet section.

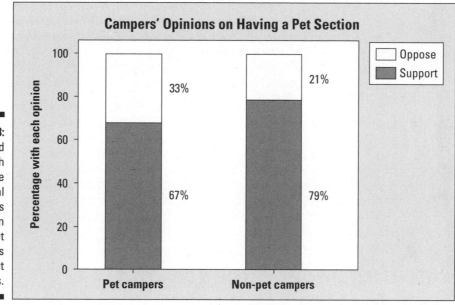

Campers' Opinions on Having a Pet Section

Figure 19-3:
Stacked
bar graph
showing the
conditional
distributions
of opinion
for pet
campers
and non–pet
campers.

Now you compare the opinions of the two groups by comparing the percentage of supporters in the pet camping group (67%) to the percentage of supporters in the non-pet camping group (79%). While both groups have a majority of supporters of the pet section, you see more of the non pet campers support the policy than pet campers (because 79% > 67%). By comparing the conditional distributions, you've found that a relationship appears to exist between opinion and pet camping, and your original hunch that non-pet campers in the area may be more in favor of a separate pet area than pet campers is correct, based on this data.

The difference in the results found in Figure 19-3 isn't as large as you may have thought by looking at the joint distribution in Figure 19-2. The conditional distribution takes into account and adjusts for the number in each group being compared, while the joint distribution puts everyone in the same boat. That's why you need conditional distributions to make fair comparisons.

When making my conclusions regarding the pet-camping data, the operative words I use are "a relationship *appears* to exist." The results of the pet camping survey are based on only your sample of 100 campers. To be able to generalize these results to the whole population of pet campers and non–pet campers in this community (which is really what you want to do), you need to take into account that these sample results will vary, and when they do vary, will they still show the same kind of difference? That's what a hypothesis test will tell you (all the details are in Chapter 14).

To conduct a hypothesis test for a relationship between two categorical variables (when each variable has only two categories, like yes/no or male/female), you either do a test for two proportions (see Chapter 15) or a Chisquare test (which is covered in my book *Statistics II For Dummies,* also published by Wiley). If one or more of your variables have more than two categories, such as Democrats/Republicans/Other, you must use the Chisquare test to test for independence in the population.

Be mindful that you may run across a report in which someone is trying to give the appearance of a stronger relationship than really exists, or trying to make a relationship less obvious by how the graphs are made. With pie charts, the sample size often is not reported, leading you to believe the results are based on a large sample when they may not be. With bar graphs, they stretch or shrink the scale to make differences appear larger or smaller, respectively. (See Chapter 6 for more information on misleading graphs of categorical data.)

Checking Independence and Describing Dependence

The main reason researchers collect data on two categorical variables is to explore possible relationships or connections between the variables. For example, if a survey finds that more females than males voted for the incumbent president in the last election, then you conclude that gender and voting outcome are related. If a relationship between two categorical variables has been found (that is, the results from the two groups are different), then statisticians say they're *dependent*.

However, if you find that the percentage of females who voted for the incumbent is the same as the percentage of males who voted for the incumbent, then the two variables (gender and voting for the incumbent) have no relationship and statisticians say those two variables are *independent*. In this section, you find out how to check for independence and describe relationships found to be dependent.

Checking for independence

Two categorical variables are *independent* if the percentages for the second variable (typically representing the results you want to compare, such as support or oppose) do not differ based on the first variable (typically representing the groups you want to compare, such as men versus women). You can check for independence with the methods that I cover in this section.

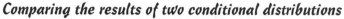
Comparing the results of two conditional distributions

Two categorical variables are *independent* if the conditional distributions are the same for all groups being compared. The variables are independent because breaking them down and comparing them by group doesn't change the results. In the election example I introduced at the beginning of "Checking Independence and Describing Dependence," independence means the conditional distribution for opinion is the same for the males and the females.

Suppose you do a survey of 200 voters to see if gender is related to whether they voted for the incumbent president, and you summarize your results in Table 19-8.

Table 19-8	Results of Election Survey		
	Voted for Incumbent President	*Didn't Vote for Incumbent President*	*Marginal Row Totals*
Males	44	66	110
Females	36	54	90
Marginal Column Totals	80	120	**Grand total = 200**

To see whether gender and voting are independent, you find the conditional distribution of voting pattern for the males and the conditional distribution of voting pattern for the females. If they're the same, you've got independence; if not, you've got dependence. These two conditional distributions have been calculated and appear in rows 1 and 2, respectively, of Table 19-9. (See the earlier section "Comparing groups with conditional distributions" for details.)

To get the numbers in Table 19-9, I started with Table 19-8 and divided the number in each cell by its marginal row total to get a proportion. Each row in Table 19-9 sums to 1 because each row represents its own conditional distribution. (If you're male, you either voted for the incumbent or you didn't — same for females.)

Row 1 of Table 19-9 shows the conditional distribution of voting pattern for males. You see 40% voted for the incumbent and 60% not. Similarly, row 2 of the table shows the conditional distribution of voting pattern for females; again, 40% voted for the incumbent and 60% did not. Because these distributions are the same, men and women voted the same way; gender and voting pattern are independent.

Table 19-9	Results of Election Survey with Conditional Distributions		
	Voted for Incumbent President	*Didn't Vote for Incumbent President*	*Total*
Males	44 ÷ 110 = 0.40	66 ÷ 110 = 0.60	1.00
Females	36 ÷ 90 = 0.40	54 ÷ 90 = 0.60	1.00

Figure 19-4 shows the conditional distributions of voting pattern for males and females using a graph called a stacked bar chart. Because the bars look exactly alike, you conclude that gender and voting pattern are independent.

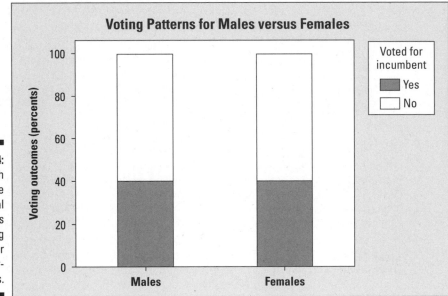

Figure 19-4:
Bar graph showing the conditional distributions of voting pattern for males versus females.

To have independence, you don't need the percentages within each bar to be 50-50 (for example, 50% males in favor and 50% males opposed). It's not the percentages within each bar (group) that have to be the same; it's the percentages across the bars (groups) that need to match (for example, 60% of males in favor and 60% of females in favor).

Instead of comparing rows of a two-way table to determine independence, you can compare the columns. In the voting example you'd be comparing the gender breakdowns for the group who voted for the incumbent to the gender breakdowns for the group who didn't vote for the incumbent. The conclusion of independence would be the same as what you found previously, although the percentages you calculate would be different.

Comparing marginal and conditional to check for independence

Another way to check for independence is to see whether the marginal distribution of voting pattern (overall) equals the conditional distribution of voting pattern for each of the gender groups (males and females). If these distributions are equal, then gender doesn't matter. Again, gender and voting pattern are independent.

Looking at the voting pattern example, you find the conditional distribution of voting pattern for the males (first bar in Figure 19-4) is 40% yes and 60% no. To find the marginal (overall) distribution of voting pattern (males and females together), take the marginal column totals in the last row of Table 19-8 (80 yes and 120 no) and divide through by 200 (the grand total). You get 80 ÷ 200 = 0.40 or 40% yes, and 120 ÷ 200 = 0.60 or 60% no. (See the section "Calculating marginal distributions" earlier in this chapter for more explanation.) The marginal distribution of overall voting pattern matches the conditional distribution of voting pattern for males, so voting pattern is independent of gender.

Here's where a small table with only two rows and two columns cuts you a break. You have to compare only one of the conditionals to the marginal because you have only two groups to compare. If the voting pattern for the males is the same as the overall voting pattern, then the same will be true for the females. To check for independence when you have more than two groups, you use a Chi-square test (discussed in my book *Statistics II For Dummies,* published by Wiley).

Describing a dependent relationship

Two categorical variables are *dependent* if the conditional distributions are different for at least two of the groups being compared. In the election example from the previous section, the groups are males and females, and the variable being compared is whether the person voted for the incumbent president.

Dependence in this case means knowing that the outcome of the first variable does affect the outcome of the second variable. In the election example, if dependence had been found, it would mean that males and females didn't have the same voting pattern for the incumbent (for example, more males voting for the incumbent than females). (Pollsters use this kind of data to help steer their campaign strategies.)

Other ways of saying two variables are dependent are to say they are related, or associated. However, statisticians don't use the term *correlation* to indicate relationships between categorical variables. The word *correlation* in this context applies to the linear relationship between two numerical variables (such as height and weight), as seen in Chapter 18. (This mistake occurs in the media all the time, and it drives us statisticians crazy!)

Here's an example to help you better understand dependence: A recent press release put out by The Ohio State University Medical Center caught my attention. The headline said that aspirin can prevent polyps in colon-cancer patients. Having had a close relative who succumbed to this disease, I was heartened at the prospect that researchers are making progress in this area and decided to look into it.

The researchers studied 635 colon-cancer patients; they randomly assigned approximately half of them to an aspirin regimen (317 people) and the other half to a placebo (fake pill) regimen (318 people). They followed the patients to see which ones developed subsequent polyps and which did not. The data from the study are summarized in Table 19-10.

Table 19-10	Summary of Aspirin and Polyps Study Results		
	Developed Subsequent Polyps	*Didn't Develop Subsequent Polyps*	*Total*
Aspirin	54 (17%)	263 (83%)	317 (100%)
Placebo	86 (27%)	232 (73%)	318 (100%)
Total	140	495	635

Comparing the results in the rows of Table 19-10 to check for independence means finding the conditional distribution of outcomes (polyps or not) for the aspirin group and comparing it to the conditional distribution of outcomes for the placebo group. Making these calculations, you find that 54 ÷ 317 = 17% of patients in the aspirin group developed polyps (the rest, 83%, did not), compared to 86 ÷ 318 = 27% of the placebo group who developed subsequent polyps (the rest, 73%, did not).

Because the percentage of patients developing polyps is much smaller for the aspirin group compared to the placebo group (17% versus 27%), a dependent relationship appears to exist between aspirin-taking and the development of subsequent polyps among the colon-cancer patients in this study. (But does it carry over to the population? You find out in the section "Projecting from sample to population" later in this chapter.)

Cautiously Interpreting Results

It's easy to get carried away when a relationship between two variables has been found; you see this happen all the time in the media. For example, a study reports that eating eggs doesn't affect your cholesterol as once thought; in the details of the report you see the study was conducted on a total of 20 men who were all in excellent health, on low-fat diets, who exercised several times a week. Ten men in good health ate two eggs a day and their cholesterol didn't change much, compared to ten men who didn't eat two eggs per day. Do these results carry over to the entire population? Can't tell — the subjects in the study don't represent the rest of us. (See Chapter 17 for the scoop on evaluating experiments.)

In this section, you see how to put the results from a two-way table into proper perspective in terms of what you can and can't say and why. This basic understanding gives you the ability to critically evaluate and make decisions about results presented to you (not all of which are correct).

Checking for legitimate cause and effect

Researchers studying two variables often look for links that indicate a cause-and-effect relationship. A *cause-and-effect relationship* between two categorical variables means as you change the value of one variable and all else remains the same, it causes a change in the second variable — for example, if being on an aspirin regimen decreases the chance of developing subsequent polyps in colon-cancer patients.

However, just because two variables are found to be related (dependent) doesn't mean they have a cause-and-effect relationship. For example, observing that people who live near power lines are more likely to visit the hospital in a year's time due to illness doesn't necessarily mean the power lines caused the illnesses.

The most effective way to conclude a cause-and-effect relationship is by conducting a well-designed experiment (where possible). All the details are laid out in Chapter 17, but I touch on the main points here. A well-designed experiment meets the following three criteria:

- ✔ It minimizes *bias* (systematic favoritism of subjects or outcomes).
- ✔ It repeats the experiment on enough subjects so the results are reliable and repeatable by another researcher.
- ✔ It controls for other variables that may affect the outcome that weren't included in the study.

In the earlier section "Describing a dependent relationship," I discuss a study involving the use of aspirin to prevent polyps in cancer patients. Because of the way the data was collected for this study, you can be confident about the conclusions drawn by the researchers; this study was a well-designed experiment, according to the criteria established in Chapter 17. To avoid problems, the researchers in this study did the following:

✔ Randomly chose who took the aspirin and who received a fake pill

✔ Had large enough sample sizes to obtain accurate information

✔ Controlled for other variables by conducting the experiment on patients in similar situations with similar backgrounds

Because their experiment was well-designed, the researchers concluded that a cause-and-effect relationship was found for the patients in this study. The next test is to see whether they can project these results to the population of all colon-cancer patients. If so, they are truly entitled to the headline "Aspirin Prevents Polyps in Colon-Cancer Patients." The next section walks you through the test.

Whether two related variables are found to be causally associated depends on how the study was conducted. A well-designed experiment is the most convincing way to establish cause and effect. In cases where an experiment would be unethical (for example, proving that smoking causes lung cancer by forcing people to smoke), a mountain of convincing observational studies (where you collect data on people who smoke and people who don't) would be needed to show that an association between two variables crosses over into a cause-and-effect relationship.

Projecting from sample to population

In the aspirin/polyps experiment discussed in the earlier section "Describing a dependent relationship," I compare the percentage of patients developing subsequent polyps for the aspirin group versus the non-aspirin group and got the results 17% and 27%, respectively. For this sample, the difference is quite large, so I'm cautiously optimistic that these results would carry over to the population of all cancer patients. But what if the numbers were closer, such as 17% and 20%? Or 17% compared to 19%? How different do the proportions have to be in order to signal a meaningful association between the two variables?

Percentages compared using data from your sample reflect relationships within your sample. However, you know that results change from sample to sample. To project these conclusions to the population of all colon-cancer patients (or any population being studied), the difference in percentages found by the sample has to be *statistically significant*. Statistical significance means even though you know results will vary, even taking that variation into account it's very unlikely the differences were due to chance. That way, the same conclusion about a relationship can be made about the whole population, not just for a particular data set.

I analyzed the data from the aspirin/polyps study using a hypothesis test for the difference of two proportions (found in Chapter 15). The proportions being compared were the proportion of patients taking aspirin who developed subsequent polyps and the proportion of patients not taking aspirin who developed subsequent polyps. Looking at these results, my *p*-value is less than 0.0024. (A *p-value* measures how likely you were to have gotten the results from your sample if the populations really had no difference; see Chapter 14 to get the scoop on *p*-values.)

Because this *p*-value is so small, the difference in proportions between the aspirin and non-aspirin groups is declared to be statistically significant, and I conclude that a relationship exists between taking aspirin and developing fewer subsequent polyps.

You can't make conclusions about relationships between variables in a population based only on the sample results in a two-way table. You must take into account the fact that results change from sample to sample. A hypothesis test gives limits for how different the sample results can be to still say the variables are independent. Beware of conclusions based only on sample data from a two-way table.

Making prudent predictions

A common goal of research (especially medical studies) is to make predictions, recommendations, and decisions after a relationship between two categorical variables is found. However, as a consumer of information, you have to be very careful when interpreting results; some studies are better designed than others.

The colon-cancer study from the previous section shows that patients who took aspirin daily had a lower chance of developing subsequent polyps (17% compared to 27% for the non-aspirin group). Because this was a well-designed experiment and the hypothesis test for generalizing to the population was significant, making predictions and recommendations for the population of colon-cancer patients based on these sample results is appropriate. They've indeed earned the headline of their press release: "Aspirin Prevents Polyps in Colon-Cancer Patients."

Resisting the urge to jump to conclusions

Try not to jump to conclusions when you hear or see a relationship being reported regarding two categorical variables. Take a minute to figure out what's really going on, even when the media wants to sweep you away with a dramatic result.

For example, as I write this, a major news network reports that men are 40% more likely to die from cancer than women. If you're a man, you may think you should panic. But when you examine the details, you find something different. Researchers found that men are much less likely to go to the doctor than women, so by the time cancer is found, it's more advanced and difficult to treat. As a result, men were more likely to die of cancer after its diagnosis. (They aren't necessarily more likely to *get* cancer; that's for a different study.) This study was meant to promote early detection as the best protection and encourage men to keep their annual checkups. The message would have been clearer had the media reported it correctly (but that's not as exciting or dramatic).

Part VI
The Part of Tens

The 5th Wave   By Rich Tennant

Advanced Studies in Probability:
The Cat-With-Attached-Open-Faced-
Jelly-Sandwich Drop.

In this part . . .

Where would a statistics book be without some statistics of its own? This part contains ten methods for being a statistically savvy sleuth and ten tips for boosting your score on a statistics exam. You can use this quick, concise reference to help critique or design a survey, detect common statistical abuses, and ace your introductory statistics course.

Chapter 20

Ten Tips for the Statistically Savvy Sleuth

In This Chapter

▶ Recognizing common statistical mistakes made by researchers and the media

▶ Avoiding mistakes when doing your statistics

*T*his book is not only about understanding the statistics that you come across in the media and in your workplace; it's even more about digging deeper to examine whether those statistics are correct, reasonable, and fair. You have to be vigilant — and a bit skeptical — to deal with today's information explosion, because many of the statistics you find are wrong or misleading, either by error or by design. If you don't critique the information you're consuming, in terms of its correctness, completeness, and fairness, who will? In this chapter, I outline ten tips for detecting common statistical mistakes made by researchers and by the media and ways to avoid making them yourself.

Pinpoint Misleading Graphs

Most graphs and charts contain great information that makes a point clearly, concisely, and fairly. However, many graphs give incorrect, mislabeled, and/or misleading information; or they simply lack important information that the reader needs to make critical decisions about what is being presented. Some of these shortcomings occur by mistake; others are incorporated by design in hopes you won't notice. If you're able to pick out problems with a graph before you contemplate any conclusions, you won't be taken in by misleading graphs.

Figure 20-1 shows examples of four important types of data displays: pie charts, bar graphs, time charts, and histograms. In this section I point out some of the ways you can be misled if these types of graphs are not made properly. (For more information on making charts and graphs correctly and identifying misleading ones, see Chapters 6 and 7.)

Pie charts

Pie charts are exactly what they sound like: circular (pie-shaped) charts that are divided into slices that represent the percentage (relative frequency) of individuals that fall into different groups. Groups represent a categorical variable, such as gender, political party, or employment status. Figure 20-1a is a pie chart showing a breakdown of voter opinions on some issue (call it Issue 1).

Here's how to sink your teeth into a pie chart and test it for quality:

✔ Check to be sure the percentages add up to 100 percent, or close to it (any round-off error should be small).

✔ Be careful when you see a slice of the pie called "other"; this is the catch-all category. If the slice for "other" is too large (larger than other slices), the pie chart is too vague. On the other extreme, pie charts with many tiny slices give you information overload.

✔ Watch for distortions that come with the three-dimensional ("exploded") pie charts, in which the slice closest to you looks larger than it really is because of the angle at which it's presented.

✔ Look for a reported total number of individuals who make up the pie chart so you can determine how big the sample was before it was divided up into slices. If the size of the data set (the number of respondents) is too small, the information isn't reliable.

Bar graphs

A bar graph is similar to a pie chart, except that instead of being in the shape of a circle that's divided up into slices, a bar graph represents each group as a bar, and the height of the bar represents the number (frequency) or percentage (relative frequency) of individuals in that group. Figure 20-1b is a relative frequency–style bar graph showing voter opinions on some issue (call it Issue 1); its results correspond with the pie chart shown in Figure 20-1a.

When examining a bar graph:

✔ Check for the sample size. If the bars represent frequencies, you find the sample size by summing them up; if the bars represent relative frequencies, you need the sample size to know how much data went into making the graph.

✔ Consider the units being represented by the height of the bars and what the results mean in terms of those units. For example, are they showing the total number of crimes, or the crime rate (also known as total number of crimes per capita)?

✔ Evaluate the starting point of the axis where the counts (or percents) are shown, and watch for the extremes: If the heights of the bars fluctuate from 200 to 300 but the counts axis starts at 0, the heights of the bars won't look much different. However, if the starting point on the counts axis is 200, you are basically chopping off the bottoms of all the bars, and what differences remain (ranging from 0 to 100) will look more dramatic than they should.

✔ Check out the range of values on the axis where the counts (or percents) are shown. If the heights of the bars range from 6 to 108 but the axis shows 0 to 500, the graph will have a great deal of white space and differences in the bars become hard to distinguish. However, if the axis goes from 5 to 110 with almost no breathing room, the bars will be stretched to the limit, making differences between groups look larger than they should.

Time charts

A time chart shows how a numerical variable changes over time (for example, stock prices, car sales, or average temperature). Figure 20-1c is an example of a time chart showing the percentage of yes voters from 2002 to 2010, in 2-year increments.

Here are some issues to watch for with time charts:

✔ Watch the scale on the vertical (quantity) axis as well as the horizontal (timeline) axis; results can be made to look more or less dramatic than they actually are by simply changing the scale.

✔ Take into account the units being portrayed by the chart and be sure they are equitable for comparison over time; for example, are dollar amounts being adjusted for inflation?

✔ Beware of people trying to explain why a trend is occurring without additional statistics to back themselves up. A time chart generally shows *what* is happening. *Why* it's happening is another story!

✔ Watch for situations in which the time axis isn't marked with equally spaced jumps. This often happens when data are missing. For example, the time axis may have equal spacing between 2001, 2002, 2005, 2006, 2008 when it should actually show empty spaces for the years in which no data are available.

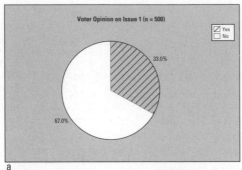

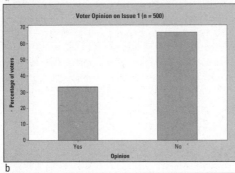

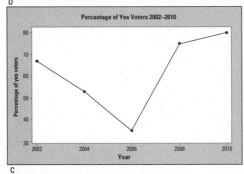

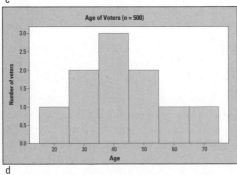

Figure 20-1:
Four types
of graphs:
a) pie chart;
b) bar graph;
c) time
chart; and
d) histogram.

Histograms

A *histogram* is a graph that breaks the sample into groups according to a numerical variable (such as age, height, weight, or income) and shows either the number of individuals (frequency) or the percentage of individuals (relative frequency) that fall into each group. Figure 20-1d is a frequency style histogram showing the ages of voters in a certain election.

Some items to watch for regarding histograms include the following:

✔ Watch the scale used for the vertical (frequency/relative frequency) axis, looking especially for results that are exaggerated or played down through the use of inappropriate scales.

✔ Check out the units on the vertical axis to see whether they report frequencies or relative frequencies; if they're relative frequencies, you need the sample size to determine how much data you're looking at.

✔ Look at the scale used for the groupings of the numerical variable on the horizontal axis. If the groups are based on small intervals (for example, 0–2, 2–4, and so on), the heights of the bars may look choppy and overly volatile. If the groups are based on large intervals (for example, 0–100, 100–200, and so on), the data may give a smoother appearance than is realistic.

Uncover Biased Data

Bias in statistics is the result of a systematic error that either overestimates or underestimates the true value. For example, if I use a ruler to measure plants and that ruler is ½-inch short, all of my results are biased; they're systematically lower than their true values.

Here are some of the most common sources of biased data:

✔ Measurement instruments may be systematically off. For example, a police officer's radar gun may say you were going 76 miles per hour when you know you were only going 72 miles per hour. Or a badly adjusted scale may always add 5 pounds to your weight.

✔ The way the study is designed can create bias. For example, a survey question that asks, "Have you *ever* disagreed with the government?" will overestimate the percentage of people who are generally unhappy with the government. (See Chapter 16 for ways to minimize bias in surveys.)

✔ The sample of individuals may not represent the population of interest — for example, examining student study habits by only going to the campus library. (See more in the section, "Identify Non-Random Samples" later in this chapter.)

✔ Researchers aren't always objective. Suppose in a drug study one group of patients is given a sugar pill and the other group is given the real drug. If the researchers know who received the real drug, they may inadvertently pay more attention to those patients to see if it's working; they may even project results onto the patients (such as saying, "I bet you're feeling better, aren't you?"). This creates a bias in favor of the drug. (See Chapter 17 for more information on setting up good experiments.)

To spot biased data, examine how the data were collected. Ask questions about the selection of the participants, how the study was conducted, what questions were used, what treatments (medications, procedures, therapy, and so on) were given (if any) and who knew about them, what measurement instruments were used and how they were calibrated, and so on. Look for systematic errors or favoritism, and if you see too much of it, ignore the results.

Search for a Margin of Error

The word *error* has a somewhat negative connotation, as if an error is something that is always avoidable. In statistics, that's not always the case. For example, a certain amount of what statisticians call *sampling error* will always occur whenever someone tries to estimate a population value using anything other than the entire population. Just the act of selecting a sample from the population means you leave out certain individuals, and that means you're not going to get the precise, exact population value. No worries, though. Remember that statistics means never having to say you're certain — you have to only get close. And if the sample is large enough, the sampling error will be small (assuming it's good data of course).

To evaluate a statistical result, you need a measure of its accuracy — typically through the margin of error. The margin of error tells you how much the researcher expects her results to vary from sample to sample. (For more information on margin of error, see Chapter 12.) When a researcher or the media fail to report the margin of error, you're left to wonder about the accuracy of the results, or worse, you just assume that everything is fine, when in many cases, it's not.

When looking at statistical results in which a number is being estimated (for example, the percentage of all Americans who think the president is doing a good job), always check for the margin of error. If it's not included, ask for it! (Or if given enough other pertinent information, you can calculate the margin of error yourself using the formulas in Chapter 13.)

Identify Non-Random Samples

If you're trying to study a population but you can only study a sample of individuals from it, how can you ensure that your sample represents the population? The most important criteria is to select your sample in a random fashion; that is, to take a *random sample*. You know a sample is random if it had the same chance of being selected as every other possible sample of the same size did. (It's like pulling names from a hat.)

Many surveys aren't based on random samples, however. For example, TV polls asking viewers to "call us with your opinion" don't represent random samples. In fact they don't represent samples at all; when you take a sample, you select individuals from the population; for call-in polls, the individuals select themselves.

Experiments (particularly medical studies) typically can't involve a random sample of individuals, for ethical reasons. You can't call someone and say, "You were chosen at random to participate in a sleep study. You'll need to come down to our lab tomorrow and stay there for two nights." Such types of experiments are conducted using subjects that volunteer to participate — they're not randomly selected first.

But even though you can't randomly select the subjects (participants) for your experiment, you can still get valid results if you incorporate the randomness in a different way — by randomly assigning the subjects to the treatment group and the control group. If the groups were assigned at random, they have a good chance of being very similar, except for what treatment they received. That way, if you do find a large enough difference in the outcomes of the groups, you can attribute those differences to the treatment, rather than to other factors.

Before making any decisions about statistical results from a survey, look to see how the sample of individuals was selected. If the sample wasn't selected randomly, take the results with a grain of salt (see Chapter 16). If you're looking at the results of an experiment, find out whether the subjects were randomly assigned to the treatment and control groups; if not, ignore the results (see Chapter 17).

Sniff Out Missing Sample Sizes

Both the quality and quantity of information is important in assessing how accurate a statistic will be. The more good data that goes into a statistic, the more accurate that statistic will be. The quality issue is tackled in the section "Uncover Biased Data" earlier in this chapter. When the quality has been established, you need to assess the accuracy of the information, and for that you need to look at how much information was collected (that is, you have to know the sample size).

Small sample sizes make results less accurate (unless your population was small to begin with). Many headlines aren't exactly what they appear to be when the details reveal a study that was based on a small sample. Perhaps even worse, many studies don't even report the sample size at all, which should lead you to be skeptical of the results. (For example, an old chewing gum ad said, "Four out of five dentists surveyed recommend [this gum] for their patients who chew gum." What if they really did ask only five dentists?)

Don't think about this too much, but according to statisticians (who are picky about precision), 4 out of 5 is much different than 4,000 out of 5,000, even though both fractions equal 80 percent. The latter represents a much more precise (repeatable) result because it's based on a much higher sample size. (Assuming it's good data, of course.) If you ever wondered how math and statistics are different, here's your answer! (Chapter 12 has more on precision.)

However, more data isn't always better data — it depends on how well the data were collected (see Chapter 16). Suppose you want to gather the opinions of city residents on a city council proposal. A small random sample with well-collected data (such as a mail survey of a small number of homes chosen at random from a city map) is much better than a large non-random sample with poorly collected data (for example, posting a Web survey on the city manager's Web site and asking for people to respond).

Always look for the sample size before making decisions about statistical information. The smaller the sample size, the less precise the information. If the sample size is missing from the article, get a copy of the full report of the study, contact the researcher, or contact the journalist who wrote the article.

Detect Misinterpreted Correlations

Everyone wants to look for connections between variables; for example, what age group is more likely to vote Democrat? If I take even more vitamin C, am I even less likely to get a cold? How does staring at the computer all day affect my eyesight? When you think of connections or associations between variables, you probably think of correlation. Yes, correlation is one of the most commonly used statistics — but it's also one of the most misunderstood and misused, especially throughout the media.

Some important points about correlation are as follows (see Chapter 18 for all the additional information):

> ✓ **The statistical definition of *correlation* (denoted by *r*) is the measure of strength and direction of the linear relationship between two numerical variables.** A correlation tells you whether the variables increase together or go in opposite directions and the extent to which the pattern is consistent across the data set.

- **The statistical term *correlation* is only used in the context of two numerical variables (such as height and weight).** It does not apply to two categorical variables (such as political party and gender).

  For example, voting pattern and gender may be related, but using the word *correlated* to describe their relationship isn't "sc" (statistically correct, get it?). You can say two categorical variables are *associated*.

- **If a strong correlation and scatterplot exist between two numerical variables, you should be able to draw a straight line through the points, and the points should lie close to the line.** If a line doesn't fit the data well, the variables likely won't have a strong correlation (*r*), and vice versa. (See Chapter 18 for information on line-fitting, also known as *linear regression*.)

  A weak correlation implies that a linear relationship doesn't exist between the two variables, but this doesn't necessarily mean the variables aren't related at all. They may have some other type of relationship besides a linear relationship. For example, bacteria multiply at an exponential rate over time (their numbers explode, doubling faster and faster).

- **Correlation doesn't automatically mean cause and effect.** For example, suppose Susan reports based on her observations that people who drink diet soda have more acne than people who don't. If you're a diet soda drinker, don't break out just yet! This correlation may be a freak coincidence that only happened to the people she observed. At most, it means more research needs to be done (beyond observation) in order to draw any connections between diet soda and acne. (Susan can read Chapter 17 to find out how to design a good experiment.)

Reveal Confounding Variables

A *confounding variable* is a variable that isn't included in a study but whose influence can affect the results and create confusing (confounding) conclusions. For example, suppose a researcher reports that eating seaweed helps you live longer, but when you examine the study, you find out that it was based on a sample of people who regularly eat seaweed in their diets and are over the age of 100. When you read the interviews of these people, you discover some of their other secrets to long life (besides eating seaweed): They slept an average of 8 hours a day, drank a lot of water, and exercised every day. So did the seaweed cause them to live longer? You can't tell, because several confounding variables (exercise, water consumption, and sleeping patterns) may also have contributed.

The best way to control for confounding variables is to conduct a well-designed experiment (see Chapter 17), which involves setting up two groups that are alike in as many ways as possible, except that one group receives a specified treatment and the other group receives a control (a fake treatment, no treatment, or a standard, non-experimental treatment). You then compare the results from the two groups, attributing any significant differences to the treatment (and to nothing else, in an ideal world).

This seaweed study wasn't a designed experiment; it was an observational study. In observational studies, no control for any variables exists; people are merely observed, and information is recorded. Observational studies are great for surveys and polls, but not for showing cause-and-effect relationships, because they don't control for confounding variables. A well-designed experiment provides much stronger evidence.

If doing an experiment is unethical (for example, showing smoking causes lung cancer by forcing half of the subjects in the experiment to smoke ten packs a day for 20 years while the other half of the subjects smoke nothing), then you must rely on mounting evidence from many observational studies over many different situations, all leading to the same result. (See Chapter 17 for all the details on designing experiments.)

Inspect the Numbers

Just because a statistic appears in the media doesn't mean it's correct. In fact, errors appear all the time (by mistake or by design), so stay on the lookout for them. Here are some tips for spotting botched numbers:

- **Make sure everything adds up to what it's reported to.** With pie charts, be sure all the percentages add up to 100 percent (subject to a small amount of rounding error).

- **Double-check even the most basic of calculations.** For example, a pie chart shows that about 83.33 percent of Americans are in favor of an issue, but the accompanying article reports "7 out of every 8" Americans are in favor of the issue. Are these statements saying the same thing? No; 7 divided by 8 is 87.5 percent — if you want 83.33 percent, it's 5 out of 6.

- **Look for the response rate of a survey; don't just be happy with the number of participants.** (The response rate is the number of people who responded divided by the total number of people surveyed times 100 percent.) If the response rate is much lower than 50 percent, the results may be biased, because you don't know what the non-respondents would have said. (See Chapter 16 for the full scoop on surveys and their response rates.)

- **Question the type of statistic used, to determine whether it's appropriate.** For example, suppose the number of crimes went up, but so did the population size. Instead of reporting the number of crimes, the media need to report the crime rate (number of crimes per capita).

Statistics are based on formulas that take the numbers you give them and crunch out what you ask them to crunch out. The formulas don't know whether the final answers are correct or not. The people behind the formulas should know better, of course. Those who don't know better will make mistakes; those who do know better might fudge the numbers anyway and hope you don't catch on. You, as a consumer of information (also known as a certified skeptic), must be the one to take action. The best policy is to ask questions.

Report Selective Reporting

You cannot credit studies in which a researcher reports his one statistically significant result but fails to mention the reports of his other 25 analyses, none of which came up significant. If you had known about all the other analyses, you may have wondered whether this one statistically significant result is truly meaningful, or simply due to chance (like the idea that a monkey typing randomly on the typewriter would eventually write Shakespeare). It's a legitimate question.

The misleading practice of analyzing data until you find something is what statisticians call *data snooping* or *data fishing*. Here's an example: Suppose Researcher Bob wants to figure out what causes first graders to argue with each other so much in school (he must not be a parent or he wouldn't even try to touch this one!). He sets up a study in which he observes a classroom of first graders every day for a month and records their every move. He gets back to his office, enters all his data, hits a button that asks the computer to perform every analysis known to man, and sits back in his chair eagerly awaiting the results. After all, with all this data he's bound to find *something*.

After poring through his results for several days, he hits pay dirt. He runs out of his office and tells his boss he's got to put out a press release saying a ground-breaking study finds that first graders argue most when 1) the day of the week ends in the letter *y* or 2) when the goldfish in their classroom aquarium swims through the hole in its sunken pirate ship. Great job, Researcher Bob! I've got a feeling that a month of watching a group of first graders took the edge off his data analysis skills.

The bottom line is that if you collect enough data and analyze it long enough, you're bound to find something, but that something may be totally meaningless or just a fluke that's not repeatable by other researchers.

How do you protect yourself against misleading results due to data fishing? Find out more details about the study, starting with how many tests were done in total, and how many of those tests were found to be non-significant. In other words, get the whole story if you can, so that you can put the significant results into perspective.

To avoid being reeled in by someone's data fishing, don't just go with the first result that you hear, especially if it makes big news and/or seems a little suspicious. Contact the researchers and ask for more information about their data, or wait to see whether other researchers can verify and replicate their results.

Expose the Anecdote

Ah, the anecdote — one of the strongest influences on public opinion and behavior ever created. And one of the least valid. An *anecdote* is a story or result based on a single person's experience or situation. For example:

- The waitress who won the lottery — twice.
- The cat that learned how to ride a bicycle.
- The woman who lost a hundred pounds in two days on the new miracle potato diet.
- The celebrity who claims to have used an over-the-counter hair color for which she is a spokesperson (yeah, right).

Anecdotes make great news; the more sensational the better. But sensational stories are outliers from the norm of life. They don't happen to most people.

You may think you're out of reach of the influence of anecdotes. But what about those times when you let one person's experience influence you? Your neighbor loves his Internet service provider, so you try it, too. Your friend had a bad experience with a certain brand of car, so you don't bother to test-drive it. Your dad knows somebody who died in a car crash because she was trapped in the car by her seat belt, so he decides never to wear his.

While some decisions are okay to make based on anecdotes, some of the more important decisions you make should be based on real statistics and real data that come from well-designed studies and careful research.

An anecdote is really a data set with a sample size of only one. You have no information to compare it to, no statistics to analyze, no possible explanations or information to go on — just a single story. Don't let anecdotes have much influence over you. Instead, rely on scientific studies and statistical information based on large random samples of individuals who represent their target populations (not just a single situation). When someone tries to persuade you by telling you an anecdote just say, "Show me the data!"

Chapter 21

Ten Surefire Exam Score Boosters

I've taught more than 40,000 students in my teaching career (don't try to guess how old I am, it's not polite!), and each student has taken at least three exams for me. That makes over 120,000 exams I've graded or had a hand in grading, and believe me, I've seen it all. I've seen excellent answers, disastrous answers, and everything in between. I've gotten notes from students in the margins asking me to go easy on them because their dog ran away and they didn't have time to study. I've seen some answers that even I couldn't figure out. I've laughed, I've cried, and I've beamed with pride at what my students have come up with in exam situations.

In this chapter, I've put together a list of ten strategies most often used by students who do well on exams. These students are not necessarily smarter than everyone else (although you do have to know your material, of course), but they are much better prepared. As a result, they are able to handle new problems and situations without getting thrown off; they make fewer little mistakes that chip away at an exam score; and they are less likely to have that deer-in-the headlights look, not being able to start a problem. They are more likely to get the right answer (or at least get partial credit) because they are good at labeling information and organizing their work. No doubt about it — preparation is the key to success on a stat exam.

You too can be a successful statistics student — or *more* successful, if you're already doing well — by following the simple strategies outlined in this chapter. Remember, every point counts, and they all add up, so let's start boosting your exam score right away!

Know What You Don't Know, and then Do Something about It

Figuring out what you know and what you don't know can be hard when you are taking a statistics class. You read the book and can understand all the examples in your notes, but you can't do your homework problems. You can answer all your roommate's statistics questions, but you can't answer your own. You walk out of an exam thinking you did well, but when you see your grade, you are shocked.

What's happening here? The bottom line is, you have to be aware of what you know and what you don't know if you want to be successful. This is a very tough skill to develop, but it's well worth it. Students often find out what they don't know the hard way — by losing points on exam questions. Mistakes are okay, we all make them — what matters is *when* you make them. If you make a mistake before the exam while you still have time to figure out what you're doing wrong, it doesn't cost you anything. If you make that same mistake on an exam, it'll cost you points.

Here's a strategy for figuring out what you know and what you don't know. Go through your lecture notes and place stars by any items from the notes that you don't understand. You can also "test" yourself, as I describe later in "Yeah-yeah trap #2," and make a list of problems that stumped you. Take your notes and list to your professor and ask him to go through the problem areas with you. Your questions will be specific enough that your professor can zoom in when he's talking with you, give you specific information and examples, and then check to make sure you understand each idea before moving on to the next item. Meeting with your professor won't take long; sometimes getting one question answered has a ripple effect and clears up other questions farther down on your list.

Leave no stone unturned when it comes to making sure you understand all the concepts, examples, formulas, notation, and homework problems before you walk into the exam. I always tell my students that 30 minutes with me has a potential of raising your grade by 10%, because I'm awfully good at explaining things and answering questions — and I'm probably better at it than any room-mate, brother-in-law, or friend who took the class four years ago with another professor. A quick office visit with your professor is well worth your time — especially if you bring a detailed list of questions with you. If for some reason your professor is not available, see if you have access to a tutor for help.

All-purpose pointers for succeeding in class

Here's some general advice my students have found helpful:

✔ I know you've heard this before, but you really are at an advantage if you go to class every day so you have a full set of notes to review. It also ensures you didn't miss any of the little things that add up to big points on an exam.

✔ Don't just write down what the professor wrote down — that's for amateurs. The professionals also write down anything else he made a big deal about but didn't write down. That's what separates the As from the Bs.

✔ Do little things to stay organized while you go through the course; you won't get overwhelmed later when it's crunch time. The day I invested 5 dollars and bought a good mechanical pencil, a good eraser, a cheap three-hole punch for my handouts, and a tiny stapler was one of the best days of my student life. Okay, it'll probably cost you 10 dollars for these items today, but trust me, it'll be worth it!

✔ Get to your know your professor and let her get to know you. Introducing yourself on the first day makes a big impression; getting face time (as well as some good help) by asking a question after class (if you have one) or stopping in during office hours never hurts. Don't worry about whether your questions are silly — it's not what level you're at now that counts; it's your desire to get to the next level and do well in the class that's important. That's what your professor wants to see.

Avoid "Yeah-Yeah" Traps

What's a "yeah-yeah" trap? It's a term I use when you get caught saying "Yeah-yeah, I got this; I know this, no problem," but then comes the exam and whoa — you didn't have it, you didn't know it, and Houston, you actually had a problem. Yeah-yeah traps are bad because they lull you into thinking you know everything, you don't have any questions, and you'll get 100% on the exam, when the truth is you still need to resolve some issues.

Although many different yeah-yeah traps exist, I point out the two most common ones in this section and help you avoid them. I call them (cleverly) *yeah-yeah trap #1* and *yeah-yeah trap #2*. Both of these traps are subtle, and they can sneak up on even the most conscientious students, so if you recognize yourself in this section, don't feel bad. Just think how many points you'll be saving yourself when you get out of "yeah-yeah" mode and into "wait a minute — here's something I need to get straightened out!" mode.

Yeah-yeah trap #1

Yeah-yeah trap #1 happens when you study by looking through your lecture notes over and over again, saying "yeah, I get that," "I understand that," and "okay, I can do that," but you don't actually try the problems from scratch totally on your own. If you understand a problem that's already been done by someone else, it only means you understand what that person did when *they* worked the problem. It doesn't say anything about whether *you* could have done it on your own in an exam situation when the pressure is on and you're staring at a blank space where your answer is supposed to be. Big difference!

I fall into yeah-yeah trap #1 too. I read through my DVR (digital video recording) manual from beginning to end, and it all made total sense to me. But a week later when I went to record a movie, I had no clue how to do it. Why not? I understood the information as I was reading along, but I didn't try to apply it for myself, and when the time came I couldn't remember how to do it.

Students always tell me, "If someone sets up the problem for me, I can always figure it out." The problem is, almost anyone can solve a problem that's already been set up. In fact, the whole point is being able to set it up, and no one is going to do that for you on an exam.

Avoid yeah-yeah trap #1 by going through your notes, pulling out a set of examples that your professor used, and writing each one on a separate piece of paper (just the problem, not the solution). Then mix up the papers and make an "exam" out of them. For each problem, try to start it by writing down just the very first step. Don't worry about finishing the problems; just concentrate on starting them. After you've done this step for all the problems, go back into your lecture notes and see if you started them right. (On the back of each problem, write down where it came from in your notes so you can check your answers faster.)

Yeah-yeah trap #2

Yeah-yeah trap #2 is even more subtle than yeah-yeah trap #1. A student comes into my office after the exam and says, "Well I worked every problem in the notes, I redid all the homework problems, I worked all the old exams you posted, and I did great on all of them; I hardly got a single problem wrong. But when I took the exam, I bombed it."

What happened? Nine out of ten times, students in yeah-yeah trap #2 did indeed work all those problems, and spent hours upon hours doing so. But whenever they got stuck and couldn't finish a problem, they peeked at the solutions (which they kept sitting right next to them), saw where they went wrong, said "yeah-yeah, that was a silly mistake — I knew that!" and continued on to finish the problem. In the end they thought they got the problems correct all by themselves, but on an exam they lost some (if not all) of the points, depending on where they originally got stuck.

So how do you avoid yeah-yeah trap #2? By making a test run under "real" exam conditions where the pressure is on. Here's how:

1. **Study as much as you need to, in whatever manner you need to, until you are ready to test your knowledge.**

2. **Sit down with a practice exam, or if one isn't available, make your own by choosing some problems from homework, your notes, or the book and shuffling them up.**

   Just like at a real exam, you also need a pencil, a calculator, and any other materials you are allowed to bring to your exam — and nothing else! Putting your book and notes away may make you feel anxious, frustrated, or exposed when you do a test run of an exam, but you really need to find out what you can do on your own before you do the real thing.

   Some teachers allow you to bring a *review sheet* (also sometimes called a *memory sheet* or — cringe — a *cheat sheet*), a sheet of paper on which you can write any helpful information you want, subject to limitations that your professor may give. If your teacher allows review sheets at tests, use one for your practice test, too.

3. **Turn on the oven timer for however long your exam is scheduled to last, and then get started.**

4. **Work as many problems as you can to the best of your ability, and when you are finished (or time runs out), put your pencil down.**

5. **When your "exam" is over, get into the lotus position and breathe in, hold it, and breathe out three times. Then look at the solutions and grade your paper the way your professor would.**

   If you couldn't start a problem, even if you just forgot one little thing and you immediately recognized it when you saw the solutions — you can't say "Yeah-yeah, I knew that; I wouldn't make that mistake on a real exam"; you have to say "No, I couldn't start it on my own. I would have gotten 0 points for that problem. I need to figure this out."

You don't get a second chance on a real exam, so when you're studying, don't be afraid to admit when you can't do a problem correctly on your own; just be glad you caught it, and figure out how to fix the problem so you'll get it right next time. Go back over it in your notes, read about it in the book, ask your professor, try more problems of the same type, or ask your study buddy to quiz you on it. Also, try to see a pattern in the type of problems that you were missing points on or getting wrong altogether. Figure out why you missed what you missed. Did you read the questions too fast, which caused you to answer them incorrectly? Was it a vocabulary or a notation issue? How did your studying align with what was on the test? And so on.

Being critical of yourself is hard, and finding out you didn't know something you thought you knew is a little scary. But if you put yourself out there and find your mistakes before they cost you points, you'll zoom in on your weaknesses, turn them into strengths, boost your knowledge, and get a higher exam score.

Make Friends with Formulas

Many students are not comfortable with formulas (unless you are a math nerd, in which case formulas make you shout for joy). That unease is understandable — I used to be intimidated by them too (formulas, that is — not math nerds). The trouble is, you really can't survive too long without eventually using a formula in a statistics class, so becoming comfortable with them right from the start is important. A formula tells you much more than how to calculate something. It shows the thinking process behind the calculations. For example, the big picture regarding standard deviation can be seen by analyzing its formula:

$$s = \sqrt{\frac{\sum_{i=1}^{n}(x_i - \bar{x})^2}{n-1}}$$

Subtracting the mean, $\bar{x}$, from a value in the data set, x_i, measures how far above or below the mean that number is. Because you don't want the positive and negative differences to cancel each other out, you square them all to make them positive (but remember that this gives you square units). Then you add them up and divide by $n - 1$, which is near to finding an average, and take the square root to get back into original units. In a general sense, you are finding something like the average distance from the mean.

Stepping back even further, you can tell from the formula that the standard deviation can't be negative, because everything is squared. You also know the smallest it can be is zero, which occurs when all the data are the same (that is, all are equal to the mean). And you see how data that is far from the mean will contribute a larger number to the standard deviation than data that is close to the mean.

And here's another perk. Because you understand the formula for standard deviation now, you know what it's really measuring: the spread of the data around the mean. So when you get an exam question saying "Measure the spread around the mean," you'll know what to do. Bam!

In order to feel comfortable about formulas, follow these tips:

✔ **Get into the right mind-set.** Think of formulas as mathematical short-hand and nothing more. All you have to do is be able to decipher them. Oftentimes you're allowed to bring a review sheet to your exam, or you'll be given a formula sheet with your exam, so you may not have to make things harder by memorizing them.

✔ **Understand every part of every formula.** In order for any formula to be useful, you have to understand all its components. For example, before you can use the formula for standard deviation, you need to know what x_i and $\bar{x}$ mean and what $\sum_{i=1}^{n}$ stands for. Otherwise it's totally useless.

✔ **Practice using formulas from day one.** Use them to verify the calculations done in lecture or in your book. If you get a different answer from what's shown, figure out what you are doing wrong. Making mistakes here is okay — you caught the problem early, and that's all that counts.

✔ **Whenever you use a formula to do a problem, write it down first and then plug in the numbers in the second step.** The more often you write down a formula, the more comfortable you will be using it on an exam. And if (heaven forbid!) you copy the formula down wrong, your instructor will be able to follow your error, which may mean some partial credit for you!

Chances are, if you've learned some formulas in your class, you're going to need to use them on your exam. Don't expect to be able to use formulas with confidence on an exam if you haven't practiced with them and written them down many, many times beforehand. Practice when the problems are easy so when they get harder you won't have to worry as much.

Make an "If-Then-How" Chart

Quarterbacks always talk about trying to get the game to "slow down" for them so they feel like they have more time to think and react. You want the same thing when you take a statistics exam. (See, you and your NFL hero really do have something in common!) The game starts slowing down for a quarterback when he begins to see patterns in the way the defense lines up against him, rather than feeling like every play brings a completely different look. Similarly for you, the exam starts to "slow down" when the problems start falling into categories as you read them, rather than each one appearing to be totally different from anything you've ever seen before.

To make this happen, many of my students find help in making what I call an if-then-how chart. An *if-then-how chart* maps out the types of problems you are likely to run into, strategies to solve them, and examples for quick reference. The basic idea of the if-then-how chart is to say "*If* the problem asks for X, *then* I solve it by doing Y, and here's *how*." An if-then-how chart contains three columns:

- ✔ ***If:*** In the *if* column, write down a succinct description of what you are asked to find or do. For example, if the problem asks you to test a claim about the population mean (see Chapter 14 for more about claims), write "Test a claim — population mean." If you are asked to give your best estimate of the population mean (Chapter 13 has the scoop on estimates), write "Estimate population mean."

  Problems are worded in different ways, because that's how the real world works. Pay attention to different wordings that in essence boil down to the same problem, and add them to the appropriate place in the *if* column where the actual problem is already listed. For example, one problem may ask you to estimate the population mean; another problem may say, "Give a range of likely values for the population mean." These questions ask for the same thing, so include both in your *if* column.

- ✔ ***Then:*** In your *then* column you write the exact statistical procedure, formula, or technique you need to solve that type of problem using the statistical lingo. For example, when your *if* column says "Test a claim — population mean," your *then* column should say "Hypothesis test for μ." When your *if* statement reads "Estimate population mean" your *then* column should read "Confidence interval for μ."

  To match strategies to situations, look carefully at how the examples in your lecture notes and your book were done and use them as your guide.

- ✔ ***How:*** In the *how* column, write an example, a formula, and/or a quick note to yourself that will spark your mind and send you off running in the right direction. Write whatever you need to feel comfortable (no one's going to see it but you, so make it your way!). For example, suppose your *if* column says "Estimate the population mean," and your *then* column says "Confidence interval — population mean." In the *how* column, you can write the formula.

Although I just took a lot of time and talking to walk you through it, making an if-then-how chart is much easier done than said. Below is an example of an entry in an if-then-how chart for the confidence interval problem I just laid out.

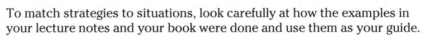

If	Then	How
Estimate the population mean (also known as range of likely values)	CI for μ	$\bar{x} \pm z^* \dfrac{\sigma}{\sqrt{n}}$

Using these three columns, fill in your if-then-how chart with each different type of problem you've covered in class. Don't write down every little example; look for patterns in the problems and boil down the number of scenarios to a doable list.

If-then-how charts should be customized to your needs, so the only way it's going to work is if you make it yourself. No two people think alike; what works for your friend may not work for you. However, it might be helpful to compare your chart with a friend's once you are both finished, to see if you've left anything out.

If you're allowed to bring a review sheet to exams, I suggest putting your if-then-how chart on one side. On the other side, write down those little nuggets of information your professor gave you in lecture but didn't write down. If you aren't allowed to have a review sheet during the exam, call me crazy, but I'll argue that you should still make one to study from. Making one really helps you sort out all the ideas so when you take the exam you'll be much more clear about what to look for and how to set up and solve problems. Lots of students come out of an exam saying they didn't even use their review sheet, and that's when you know you've done a good job putting one together: When it went on the sheet, it went into your mind!

Figure Out What the Question Is Asking

Students often tell me that they don't understand what a problem is asking for. That's the million dollar question, isn't it? And it's not a trivial matter. Oftentimes the actual question is embedded somewhere in the language of the problem; it isn't usually as clear as: "Find the mean of this data set."

For example, a question may ask you to "interpret" a statistical result. What does "interpret" really mean? To most professors the word "interpret" means to explain in words that a nonstatistician would understand.

Suppose you are given some computer output analyzing number of crimes and number of police officers, and you are asked to interpret the correlation between them. First you pick off the number from the output that represents the correlation (say it's –0.85); then you talk about its important features in language that is easy for others to understand. The answer I would like to see on an exam goes something like this: "The correlation between number of police officers and number of crimes is –0.85; they have a strong negative linear relationship. As the number of police officers increases, number of crimes decreases."

If you know what the problem is asking for, you have a better chance of actually solving it. You'll gain confidence when you know what you are supposed to do. On the flip side, if you don't know what the problem is asking, even starting it will be very hard. Your anxiety will go up, which can affect your ability to work other problems as well. So how do you boil down a problem to figure out exactly what it's asking for? Here are some tips to follow:

- ✓ **Check the very last sentence of the problem — that's usually where the question is located.** Rather than reading the entire problem a second (and third and fourth) time and getting yourself all worked up, just read it once and then focus on the end of the problem.

- ✓ **Practice boiling down questions ahead of time.** Look at all the examples from your lecture notes, your homework problems, and problems in your textbook and try to figure out what each problem is asking for. Eventually you'll start to see patterns in the way problems are worded, and you'll get better at figuring out what they are really asking for.

- ✓ **Ask your professor what clues you should look for, and bring example problems with you.** She will be impressed because you are trying to figure out the big picture, and oh, how professors love those "big picture" questions! And after she helps you, you can add those to your if-then-how chart (see "Make an 'If-Then-How' Chart").

- ✓ **Translate the wording of the problem into a statistical statement.** This involves labeling not only what you are given (as discussed in the next section), but also what you want to find.

  For example, Professor Barb wants to give 20 percent of her students an A on her statistics exam; your job is to find the cutoff exam score for an A, and this translates to "find the score representing the 80th percentile."

Label What You're Given

Many students try to work problems by pushing around numbers that are given in the problem. This approach may work with easy problems, but everyone hits the wall at some point and needs more support to solve harder problems. You'll benefit from getting into the habit of labeling everything properly — labeling is the critical connection between the *if* column and the *then* column in your if-then-how chart (described earlier in this chapter). You may read a problem and know what you need to do, but without understanding how to use what you're given in the problem, you won't be able to solve it correctly. To really understand the numbers the problem gives you, take each one and write down what it stands for.

Suppose you're given the following problem to solve: "You want to use the size of a house in a certain city (in square feet) to predict its price (in thousands). You collect data on 100 randomly selected homes that have recently been sold. You find the mean price is $219,100 with standard deviation of $60,100, and you know the mean size is 1,993 square feet, with standard deviation of 349 square feet. You find the correlation between size and price for these homes is +0.90. Find the best-fitting regression line that you can use to predict house price using size."

Your first step is labeling everything. Knowing you use size to predict price, you figure size must be the x variable and price must be the y variable. You then label the means $\bar{x} = 1,993$ (square feet) and $\bar{y} = 219.1$ (in thousands) respectively; the standard deviations are labeled $s_x = 349$ (square feet) and $s_y = 60.1$ (in thousands), respectively, and the correlation is labeled $r = 0.90$. The sample size is $n = 100$. Now you can plug your numbers into the right formulas. (See Chapter 18 regarding correlation and regression.)

When you know you have to work with a regression line and that formulas are involved, having all the given information organized and labeled, ready to go, is very comforting. It's one less thing to think about. (The problem in this particular example is solved in the section "Make the Connection and Solve the Problem.") If that example doesn't convince you, here are six more reasons to label what you are given in a problem:

- ✔ **Labeling allows you to check your work more easily.** When you go back to check your work (as I advise in the section "Do the Math — Twice"), you'll quickly see what you were thinking when you did the problem the first time.

- ✔ **Your professor will be impressed.** He will see your labels and realize you at least know what the given information stands for. That way if your calculations go haywire, you still have a chance for partial credit.

- ✔ **Labeling saves time.** I know that writing down more information seems like a strange way to save time, but by labeling all the items, you can pull out the info you need in a flash.

  For example, suppose you need to do a 95% confidence interval for the population mean (using what you know from Chapter 13) and you're told that the sample mean is 60, the population standard deviation is 10, and the sample size is 200. You know the formula has to involve $\bar{x}$, σ, and n, and you see one that does:

  $$\bar{x} \pm z * \frac{\sigma}{\sqrt{n}}$$

  Because you've already labeled everything, you just grab what you need, put it into the formula, throw in a z^*-value of 1.96 (the critical value corresponding to a 95% confidence level), and crunch it out to get the answer:

  $$60 \pm 1.96 \frac{10}{\sqrt{200}} = 60 \pm 1.39$$

- ✔ **Labels keep your mind organized.** You are less likely to get buried in calculations and forget what you're doing if your work involves symbols and not just numbers. By sorting out the information you are given, you're less likely to resort to reading the problem over and over again, raising your anxiety level each time.

- ✔ **You use the labels to figure out which formula or technique you need to use to solve a problem.** For example, if you think you need a hypothesis test but no claim is made about the population mean, hold up. You may need a confidence interval instead; this realization saves you precious time because you won't be spinning your wheels in the wrong direction. Labels help you quickly narrow down your options.

- ✔ **Labeling helps you resist the urge to just write down numbers and push them around on the paper.** More often than not, number-pushing leads to wrong answers and less (if any) partial credit if your answer is wrong. Your professor may not be able to follow you, or just doesn't want to spend all that time trying to figure it out (sorry to say, but this happens sometimes).

Labeling saves you anxiety, time, and points when you take your exam. But in order to be successful on exam day, you need to start this practice early on, while the problems are easy to do. Don't expect to suddenly be able to sort out the information on exam day if you never did it before; it's not gonna happen. Make it your habit right away and you won't freak out when you see a new problem. You'll at least be able to break it down into smaller chunks, which always helps.

Draw a Picture

You've heard the expression "A picture is worth a thousand words." As a statistics professor, I say, "A picture is worth a thousand points (or at least half the points on a given problem)." When the given information and/or the question being asked can be expressed in a picture form, you should do it. Here's why:

- ✔ **A picture can help you see what's going on in the problem.** For example, if you know exam scores have a normal distribution with mean 75 and standard deviation 5 (see Chapter 9 for more about normal distribution), you draw a bell-shaped curve, marking off the mean in the center and three standard deviations on each side. You can now visualize the scenario you're dealing with.

- ✔ **You can use the drawing to help figure out what you are trying to find.** For example, if you need to know the probability that Bob scored more than 70 points on the exam, you shade in the area to the right of 70 on your drawing, and you're on your way.

✔ **Your professor knows that you understand the basics of the problem, increasing your chance for partial credit.** On the other hand, someone who got the problem wrong doesn't get much sympathy if the professor knows drawing a simple picture would have avoided the whole problem.

✔ **Students who draw pictures tend to get more problems correct than students who don't.** Without a picture you can easily lose track of what's needed, and make mistakes like finding $P(X < 70)$ instead of $P(X > 70)$, for example. Also, checking for and spotting errors before you turn in your exam is easier if you have a picture to look at.

Drawing a picture may seem like a waste of valuable time on an exam, but it's actually a time-saver because it gets you going in the right direction, keeps you focused throughout the problem, and helps ensure you answer the right question. Drawing a picture can also help you analyze your final numerical answer and either confirm you've got it right, or quickly a spot and fix an error and save yourself some points. (Be sure to draw pictures while studying so they come naturally during an exam.)

Make the Connection and Solve the Problem

When you've figured out what the problem is asking, you have everything labeled, and you have your pictures drawn, it's time to solve the problem. After doing the prep work, nine times out of ten you'll remember a technique you learned from class, a formula that contains the items you've labeled, and/ or an example you worked through. Use or remember your if-then-how chart and you'll be on your way. (See "Make an 'If-Then-How' Chart" if you need more info.)

Breaking down a problem means having less to think about at each step, and in a stressful exam situation where you may forget your own name, that's a real plus! (This strategy reminds me of the saying, "How do you eat an elephant? One bite at a time.")

In the example of using size of a home to predict its price (see the earlier section "Label What You're Given"), you know the mean and standard deviation of size, the mean and standard deviation of price, and the correlation between them; and you've labeled them all. The question asks you to find the equation of the best-fitting regression line to predict price based on size of the home; you know that means find the equation $y = a + bx$ where x = size (square feet) and y = price (thousands of dollars), b is the slope of the regression line, and a is the y-intercept. (Flip to Chapter 18 for more about this formula.)

Now you recognize what to do — you have to find a and b. You remember (or can find) that those formulas are $b = r\frac{s_y}{s_x}$ and $a = \overline{y} - b\overline{x}$. Grab the numbers you've labeled ($\overline{x} = 1,993$; $s_x = 349$; $\overline{y} = 219.1$; $s_y = 60.1$; and $r = 0.90$), put them into the formulas, and solve (sounds like a commercial for a frozen dinner doesn't it?). You find the slope is $b = 0.90\frac{60.1}{349} = 0.155$ and the y-intercept is $a = 219.1 - 0.155(1,993) = -89.82$, so the equation of the best-fitting regression line is $y = -89.82 + 0.155x$. (See Chapter 18 for the details of regression.)

Do the Math — Twice

I can still remember some of the struggles I had way back in high school algebra. For the longest time 3 times 2 was equal to 5 for me; this mistake (and others like it) caused me to miss a handful of points on every exam and homework assignment, and I just could not get past it. One day I decided I'd had enough of losing points here and there for silly errors, and I did something about it. From that day on, I wrote out all of my work, step by step, and resisted the urge to do steps in my head. When I got my final answer, instead of moving on, I went back and checked every step, and I did so with the mindset that a mistake had probably slipped in somewhere and it was my job to find it before anyone else did.

This approach forced me to look at each step with fresh eyes, as if I were grading someone else's paper. I caught more mistakes because I never skipped over a step without bothering to check it. I finally stopped thinking 3 times 2 was 5 because I caught myself in the act enough times. My exam grades went up, just because I started checking things more carefully. It reminds me of the carpenter's saying, "Measure twice, cut once." They waste a lot less wood that way.

Every time you find and fix a mistake before you turn in your exam, you're getting a handful of points back for yourself. Find your errors before your professor does, and you'll be amazed how those points add up. However, remember that time is not unlimited on an exam, so try to get the problems right the first time. Labeling everything, drawing pictures, writing down formulas, and showing all your work will definitely help!

Analyze Your Answers

A very prominent statistician I know has a framed piece of paper on his office wall. It's a page of an exam he took way back when he was a student. It's got a big red circle around one of his answers, which happens to be the number 2. Why was writing the number 2 for an answer such a problem? Because the question asked him to find a probability, and probabilities are always between 0 and 1. As a result, he didn't get any points for that problem, not even partial credit. In fact, I'll bet his professor wanted to give him negative points for making such a mistake. (They really don't like it when you totally miss the boat.)

Always take the time to check your final answer to see if it makes sense. A negative standard deviation, a probability more than 1, or a correlation of −121.23 is not going to go over well with your professor, and it will not be treated like a simple math error. It will be treated as a fundamental error in not knowing (or perhaps caring) what the result should look like.

If you know an answer you got can't possibly be right, but you cannot for the life of you figure out where you went wrong, don't waste any more time on it. Just write a note in the margin that says you know your answer can't be right but you can't figure out your error. This helps separate you from the regular Joe who found a probability of 10,524.31 (yes, I've seen it) and merrily moved on.

By the way, you may be wondering why this world-class statistician still keeps this exam page framed on his office wall. He says it's to keep him humble. Learn from his example and never move on to the next problem without stepping back and saying "does this answer even make sense?"

Appendix

Tables for Reference

• •

*T*his appendix includes tables for finding probabilities and/or critical values for the three distributions used in this book: the *Z*-distribution (standard normal), the *t*-distribution, and the binomial distribution.

The Z-Table

Table A-1 shows less-than-or-equal-to probabilities for the *Z*-distribution; that is, $p(Z \leq z)$ for a given *z*-value. (See Chapter 9 for calculating *z*-values for a normal distribution; see Chapter 11 for calculating *z*-values for a sampling distribution.) To use Table A-1, do the following:

1. **Determine the *z*-value for your particular problem.**

   The *z*-value should have one leading digit before the decimal point (positive, negative, or zero) and two digits after the decimal point; for example *z* = 1.28, –2.69, or 0.13.

2. **Find the row of the table corresponding to the leading digit and first digit after the decimal point.**

   For example, if your *z*-value is 1.28, look in the "1.2" row; if *z* = –1.28, look in the "–1.2" row.

3. **Find the column corresponding to the second digit after the decimal point.**

   For example, if your *z*-value is 1.28 or –1.28, look in the ".08" column.

4. **Intersect the row and column from Steps 2 and 3.**

   This number is the probability that *Z* is less than or equal to your *z*-value. In other words, you've found $p(Z \leq z)$. For example, if *z* = 1.28, you see $p(Z \leq 1.28) = 0.8997$. For *z* = –1.28, you see $p(Z \leq -1.28) = 0.1003$.

Table A-1   The *Z*-Table

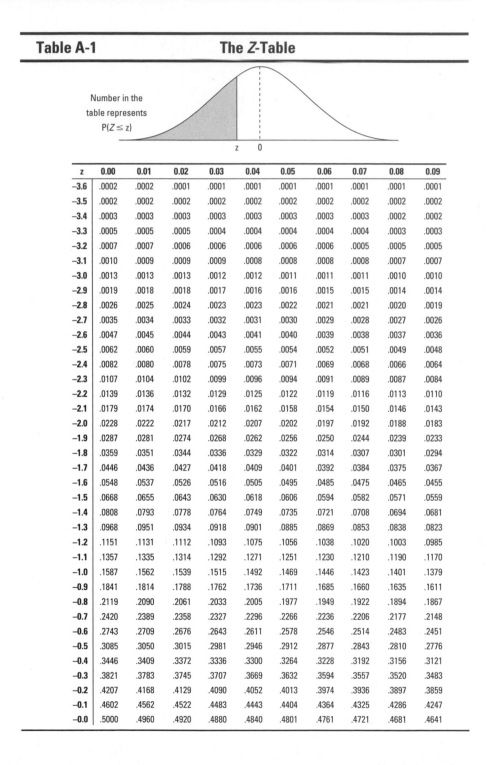

Number in the table represents $P(Z \le z)$

z	0.00	0.01	0.02	0.03	0.04	0.05	0.06	0.07	0.08	0.09
−3.6	.0002	.0002	.0001	.0001	.0001	.0001	.0001	.0001	.0001	.0001
−3.5	.0002	.0002	.0002	.0002	.0002	.0002	.0002	.0002	.0002	.0002
−3.4	.0003	.0003	.0003	.0003	.0003	.0003	.0003	.0003	.0002	.0002
−3.3	.0005	.0005	.0005	.0004	.0004	.0004	.0004	.0004	.0003	.0003
−3.2	.0007	.0007	.0006	.0006	.0006	.0006	.0006	.0005	.0005	.0005
−3.1	.0010	.0009	.0009	.0009	.0008	.0008	.0008	.0008	.0007	.0007
−3.0	.0013	.0013	.0013	.0012	.0012	.0011	.0011	.0011	.0010	.0010
−2.9	.0019	.0018	.0018	.0017	.0016	.0016	.0015	.0015	.0014	.0014
−2.8	.0026	.0025	.0024	.0023	.0023	.0022	.0021	.0021	.0020	.0019
−2.7	.0035	.0034	.0033	.0032	.0031	.0030	.0029	.0028	.0027	.0026
−2.6	.0047	.0045	.0044	.0043	.0041	.0040	.0039	.0038	.0037	.0036
−2.5	.0062	.0060	.0059	.0057	.0055	.0054	.0052	.0051	.0049	.0048
−2.4	.0082	.0080	.0078	.0075	.0073	.0071	.0069	.0068	.0066	.0064
−2.3	.0107	.0104	.0102	.0099	.0096	.0094	.0091	.0089	.0087	.0084
−2.2	.0139	.0136	.0132	.0129	.0125	.0122	.0119	.0116	.0113	.0110
−2.1	.0179	.0174	.0170	.0166	.0162	.0158	.0154	.0150	.0146	.0143
−2.0	.0228	.0222	.0217	.0212	.0207	.0202	.0197	.0192	.0188	.0183
−1.9	.0287	.0281	.0274	.0268	.0262	.0256	.0250	.0244	.0239	.0233
−1.8	.0359	.0351	.0344	.0336	.0329	.0322	.0314	.0307	.0301	.0294
−1.7	.0446	.0436	.0427	.0418	.0409	.0401	.0392	.0384	.0375	.0367
−1.6	.0548	.0537	.0526	.0516	.0505	.0495	.0485	.0475	.0465	.0455
−1.5	.0668	.0655	.0643	.0630	.0618	.0606	.0594	.0582	.0571	.0559
−1.4	.0808	.0793	.0778	.0764	.0749	.0735	.0721	.0708	.0694	.0681
−1.3	.0968	.0951	.0934	.0918	.0901	.0885	.0869	.0853	.0838	.0823
−1.2	.1151	.1131	.1112	.1093	.1075	.1056	.1038	.1020	.1003	.0985
−1.1	.1357	.1335	.1314	.1292	.1271	.1251	.1230	.1210	.1190	.1170
−1.0	.1587	.1562	.1539	.1515	.1492	.1469	.1446	.1423	.1401	.1379
−0.9	.1841	.1814	.1788	.1762	.1736	.1711	.1685	.1660	.1635	.1611
−0.8	.2119	.2090	.2061	.2033	.2005	.1977	.1949	.1922	.1894	.1867
−0.7	.2420	.2389	.2358	.2327	.2296	.2266	.2236	.2206	.2177	.2148
−0.6	.2743	.2709	.2676	.2643	.2611	.2578	.2546	.2514	.2483	.2451
−0.5	.3085	.3050	.3015	.2981	.2946	.2912	.2877	.2843	.2810	.2776
−0.4	.3446	.3409	.3372	.3336	.3300	.3264	.3228	.3192	.3156	.3121
−0.3	.3821	.3783	.3745	.3707	.3669	.3632	.3594	.3557	.3520	.3483
−0.2	.4207	.4168	.4129	.4090	.4052	.4013	.3974	.3936	.3897	.3859
−0.1	.4602	.4562	.4522	.4483	.4443	.4404	.4364	.4325	.4286	.4247
−0.0	.5000	.4960	.4920	.4880	.4840	.4801	.4761	.4721	.4681	.4641

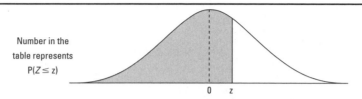

Number in the
table represents
$P(Z \le z)$

z	0.00	0.01	0.02	0.03	0.04	0.05	0.06	0.07	0.08	0.09
0.0	.5000	.5040	.5080	.5120	.5160	.5199	.5239	.5279	.5319	.5359
0.1	.5398	.5438	.5478	.5517	.5557	.5596	.5636	.5675	.5714	.5753
0.2	.5793	.5832	.5871	.5910	.5948	.5987	.6026	.6064	.6103	.6141
0.3	.6179	.6217	.6255	.6293	.6331	.6368	.6406	.6443	.6480	.6517
0.4	.6554	.6591	.6628	.6664	.6700	.6736	.6772	.6808	.6844	.6879
0.5	.6915	.6950	.6985	.7019	.7054	.7088	.7123	.7157	.7190	.7224
0.6	.7257	.7291	.7324	.7357	.7389	.7422	.7454	.7486	.7517	.7549
0.7	.7580	.7611	.7642	.7673	.7704	.7734	.7764	.7794	.7823	.7852
0.8	.7881	.7910	.7939	.7967	.7995	.8023	.8051	.8078	.8106	.8133
0.9	.8159	.8186	.8212	.8238	.8264	.8289	.8315	.8340	.8365	.8389
1.0	.8413	.8438	.8461	.8485	.8508	.8531	.8554	.8577	.8599	.8621
1.1	.8643	.8665	.8686	.8708	.8729	.8749	.8770	.8790	.8810	.8830
1.2	.8849	.8869	.8888	.8907	.8925	.8944	.8962	.8980	.8997	.9015
1.3	.9032	.9049	.9066	.9082	.9099	.9115	.9131	.9147	.9162	.9177
1.4	.9192	.9207	.9222	.9236	.9251	.9265	.9279	.9292	.9306	.9319
1.5	.9332	.9345	.9357	.9370	.9382	.9394	.9406	.9418	.9429	.9441
1.6	.9452	.9463	.9474	.9484	.9495	.9505	.9515	.9525	.9535	.9545
1.7	.9554	.9564	.9573	.9582	.9591	.9599	.9608	.9616	.9625	.9633
1.8	.9641	.9649	.9656	.9664	.9671	.9678	.9686	.9693	.9699	.9706
1.9	.9713	.9719	.9726	.9732	.9738	.9744	.9750	.9756	.9761	.9767
2.0	.9772	.9778	.9783	.9788	.9793	.9798	.9803	.9808	.9812	.9817
2.1	.9821	.9826	.9830	.9834	.9838	.9842	.9846	.9850	.9854	.9857
2.2	.9861	.9864	.9868	.9871	.9875	.9878	.9881	.9884	.9887	.9890
2.3	.9893	.9896	.9898	.9901	.9904	.9906	.9909	.9911	.9913	.9916
2.4	.9918	.9920	.9922	.9925	.9927	.9929	.9931	.9932	.9934	.9936
2.5	.9938	.9940	.9941	.9943	.9945	.9946	.9948	.9949	.9951	.9952
2.6	.9953	.9955	.9956	.9957	.9959	.9960	.9961	.9962	.9963	.9964
2.7	.9965	.9966	.9967	.9968	.9969	.9970	.9971	.9972	.9973	.9974
2.8	.9974	.9975	.9976	.9977	.9977	.9978	.9979	.9979	.9980	.9981
2.9	.9981	.9982	.9982	.9983	.9984	.9984	.9985	.9985	.9986	.9986
3.0	.9987	.9987	.9987	.9988	.9988	.9989	.9989	.9989	.9990	.9990
3.1	.9990	.9991	.9991	.9991	.9992	.9992	.9992	.9992	.9993	.9993
3.2	.9993	.9993	.9994	.9994	.9994	.9994	.9994	.9995	.9995	.9995
3.3	.9995	.9995	.9995	.9996	.9996	.9996	.9996	.9996	.9996	.9997
3.4	.9997	.9997	.9997	.9997	.9997	.9997	.9997	.9997	.9997	.9998
3.5	.9998	.9998	.9998	.9998	.9998	.9998	.9998	.9998	.9998	.9998
3.6	.9998	.9998	.9999	.9999	.9999	.9999	.9999	.9999	.9999	.9999

The t-Table

Table A-2 shows right-tail probabilities for selected t-distributions (see Chapter 10 for more on the t-distribution).

Follow these steps to use Table A-2 to find right-tail probabilities and p-values for hypothesis tests involving t (see Chapter 15):

1. **Find the t-value for which you want the right-tail probability (call it t), and find the sample size (for example, n).**

2. **Find the row corresponding to the degrees of freedom (df) for your problem (for example, $n - 1$). Go across that row to find the two t-values between which your t falls.**

   For example, if your t is 1.60 and your n is 7, you look in the row for $df = 7 - 1 = 6$. Across that row you find your t lies between t-values 1.44 and 1.94.

3. **Go to the top of the columns containing the two t-values from Step 2.**

   The right-tail (greater-than) probability for your t-value is somewhere between the two values at the top of these columns. For example, your $t = 1.60$ is between t-values 1.44 and 1.94 ($df = 6$); so the right tail probability for your t is between 0.10 (column heading for $t = 1.44$); and 0.05 (column heading for $t = 1.94$).

The row near the bottom with Z in the df column gives right-tail (greater-than) probabilities from the Z-distribution (Chapter 10 shows Z's relationship with t).

Use Table A-2 to find t^*-values (critical values) for a confidence interval involving t (see Chapter 13):

1. **Determine the confidence level you need (as a percentage).**

2. **Determine the sample size (for example, n).**

3. **Look at the bottom row of the table where the percentages are shown. Find your % confidence level there.**

4. **Intersect this column with the row representing your degrees of freedom (df).** This is the t-value you need for your confidence interval.

   For example, a 95% confidence interval with df=6 has t^*=2.45. (Find 95% on the last line and go up to row 6.)

Table A-2 The *t*-Table

Numbers in each row of the table are values on a *t*-distribution with
(*df*) degrees of freedom for selected right-tail (greater-than) probabilities (*p*).

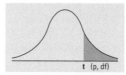

df/p	0.40	0.25	0.10	0.05	0.025	0.01	0.005	0.0005
1	0.324920	1.000000	3.077684	6.313752	12.70620	31.82052	63.65674	636.6192
2	0.288675	0.816497	1.885618	2.919986	4.30265	6.96456	9.92484	31.5991
3	0.276671	0.764892	1.637744	2.353363	3.18245	4.54070	5.84091	12.9240
4	0270722	0.740697	1.533206	2.131847	2.77645	3.74695	4.60409	8.6103
5	0.267181	0.726687	1.475884	2.015048	2.57058	3.36493	4.03214	6.8688
6	0.264835	0.717558	1.439756	1.943180	2.44691	3.14267	3.70743	5.9588
7	0.263167	0.711142	1.414924	1.894579	2.36462	2.99795	3.49948	5.4079
8	0.261921	0.706387	1.396815	1.859548	2.30600	2.89646	3.35539	5.0413
9	0.260955	0.702722	1.383029	1.833113	2.26216	2.82144	3.24984	4.7809
10	0260185	0.699812	1.372184	1.812461	2.22814	2.76377	3.16927	4.5869
11	0259556	0.697445	1.363430	1.795885	2.20099	2.71808	3.10581	4.4370
12	0259033	0.695483	1.356217	1.782288	2.17881	2.68100	3.05454	43178
13	0.258591	0.693829	1.350171	1.770933	2.16037	2.65031	3.01228	4.2208
14	0.258213	0.692417	1.345030	1.761310	2.14479	2.62449	2.97684	4.1405
15	0.257885	0.691197	1.340606	1.753050	2.13145	2.60248	2.94671	4.0728
16	0257599	0.690132	1.336757	1.745884	2.11991	2.58349	2.92078	4.0150
17	0.257347	0.689195	1.333379	1.739607	2.10982	2.56693	2.89823	3.9651
18	0.257123	0.688364	1.330391	1.734064	2.10092	2.55238	2.87844	3.9216
19	0.256923	0.687621	1.327728	1.729133	2.09302	2.53948	2.86093	3.8834
20	0.256743	0.686954	1.325341	1.724718	2.08596	2.52798	2.84534	3.8495
21	0.256580	0.686352	1.323188	1.720743	2.07961	2.51765	2.83136	3.8193
22	0256432	0.685805	1.321237	1.717144	2.07387	2.50832	2.81876	3.7921
23	0256297	0.685306	1.319460	1.713872	2.06866	2.49987	2.80734	3.7676
24	0.256173	0.684850	1.317836	1.710882	2.06390	2.49216	2.79694	3.7454
25	0.256060	0.684430	1.316345	1.708141	2.05954	2.48511	2.78744	3.7251
26	0.255955	0.684043	1.314972	1.705618	2.05553	2.47863	2.77871	3.7066
27	0.255858	0.683685	1.313703	1.703288	2.05183	2.47266	2.77068	3.6896
28	0.255768	0.683353	1.312527	1.701131	2.04841	2.46714	2.76326	3.6739
29	0.255684	0.683044	1.311434	1.699127	2.04523	2.46202	2.75639	3.6594
30	0.255605	0.682756	1.310415	1.697261	2.04227	2.45726	2.75000	3.6460
z	0.253347	0.674490	1.281552	1.644854	1.95996	2.32635	2.57583	3.2905
CI	———	———	80%	90%	95%	98%	99%	99.9%

The Binomial Table

Table A-3 shows probabilities for the binomial distribution (see Chapter 8).

To use Table A-3, do the following:

1. **Find these three numbers for your particular problem:**
   - The sample size, n
   - The probability of success, p
   - The x-value for which you want $p(X = x)$

2. **Find the section of Table A-3 that's devoted to your n.**

3. **Look at the row for your x-value and the column for your p.**

4. **Intersect that row and column.** You have found $p(X = x)$.

5. **To get the probability of being less than, greater than, greater than or equal to, less than or equal to, or between two values of X, you add the appropriate values of Table A-3 using the steps found in Chapter 8.**

For example, if $n=10$, $p=0.6$, and you want $p(X=9)$, go to the $n=10$ section, the $x=9$ row, and the $p=0.6$ column to find 0.04.

Table A-3 — The Binomial Table

Numbers in the table represent $p(X=x)$ for a binomial distribution with n trials and probability of success p.

Binomial probabilities:

$$\binom{n}{x} p^x (1-p)^{n-x}$$

n	x	0.1	0.2	0.25	0.3	0.4	0.5	0.6	0.7	0.75	0.8	0.9
1	0	0.900	0.800	0.750	0.700	0.600	0.500	0.400	0.300	0.250	0.200	0.100
	1	0.100	0.200	0.250	0.300	0.400	0.500	0.600	0.700	0.750	0.800	0.900
2	0	0.810	0.640	0.563	0.490	0.360	0.250	0.160	0.090	0.063	0.040	0.010
	1	0.180	0.320	0.375	0.420	0.480	0.500	0.480	0.420	0.375	0.320	0.180
	2	0.010	0.040	0.063	0.090	0.160	0.250	0.360	0.490	0.563	0.640	0.810
3	0	0.729	0.512	0.422	0.343	0.216	0.125	0.064	0.027	0.016	0.008	0.001
	1	0.243	0.384	0.422	0.441	0.432	0.375	0.288	0.189	0.141	0.096	0.027
	2	0.027	0.096	0.141	0.189	0.288	0.375	0.432	0.441	0.422	0.384	0.243
	3	0.001	0.008	0.016	0.027	0.064	0.125	0.216	0.343	0.422	0.512	0.729
4	0	0.656	0.410	0.316	0.240	0.130	0.063	0.026	0.008	0.004	0.002	0.000
	1	0.292	0.410	0.422	0.412	0.346	0.250	0.154	0.076	0.047	0.026	0.004
	2	0.049	0.154	0.211	0.265	0.346	0.375	0.346	0.265	0.211	0.154	0.049
	3	0.004	0.026	0.047	0.076	0.154	0.250	0.346	0.412	0.422	0.410	0.292
	4	0.000	0.002	0.004	0.008	0.026	0.063	0.130	0.240	0.316	0.410	0.656
5	0	0.590	0.328	0.237	0.168	0.078	0.031	0.010	0.002	0.001	0.000	0.000
	1	0.328	0.410	0.396	0.360	0.259	0.156	0.077	0.028	0.015	0.006	0.000
	2	0.073	0.205	0.264	0.309	0.346	0.312	0.230	0.132	0.088	0.051	0.008
	3	0.008	0.051	0.088	0.132	0.230	0.312	0.346	0.309	0.264	0.205	0.073
	4	0.000	0.006	0.015	0.028	0.077	0.156	0.259	0.360	0.396	0.410	0.328
	5	0.000	0.000	0.001	0.002	0.010	0.031	0.078	0.168	0.237	0.328	0.590
6	0	0.531	0.262	0.178	0.118	0.047	0.016	0.004	0.001	0.000	0.000	0.000
	1	0.354	0.393	0.356	0.303	0.187	0.094	0.037	0.010	0.004	0.002	0.000
	2	0.098	0.246	0.297	0.324	0.311	0.234	0.138	0.060	0.033	0.015	0.001
	3	0.015	0.082	0.132	0.185	0.276	0.313	0.276	0.185	0.132	0.082	0.015
	4	0.001	0.015	0.033	0.060	0.138	0.234	0.311	0.324	0.297	0.246	0.098
	5	0.000	0.002	0.004	0.010	0.037	0.094	0.187	0.303	0.356	0.393	0.354
	6	0.000	0.000	0.000	0.001	0.004	0.016	0.047	0.118	0.178	0.262	0.531
7	0	0.478	0.210	0.133	0.082	0.028	0.008	0.002	0.000	0.000	0.000	0.000
	1	0.372	0.367	0.311	0.247	0.131	0.055	0.017	0.004	0.001	0.000	0.000
	2	0.124	0.275	0.311	0.318	0.261	0.164	0.077	0.025	0.012	0.004	0.000
	3	0.023	0.115	0.173	0.227	0.290	0.273	0.194	0.097	0.058	0.029	0.003
	4	0.003	0.029	0.058	0.097	0.194	0.273	0.290	0.227	0.173	0.115	0.023
	5	0.000	0.004	0.012	0.025	0.077	0.164	0.261	0.318	0.311	0.275	0.124
	6	0.000	0.000	0.001	0.004	0.017	0.055	0.131	0.247	0.311	0.367	0.372
	7	0.000	0.000	0.000	0.000	0.002	0.008	0.028	0.082	0.133	0.210	0.478

(continued)

Table A-3 *(continued)*

Numbers in the table represent $p(X=x)$ for a binomial distribution with n trials and probability of success p.

Binomial probabilities:

$$\binom{n}{x} p^x (1-p)^{n-x}$$

							p					
n	x	0.1	0.2	0.25	0.3	0.4	0.5	0.6	0.7	0.75	0.8	0.9
8	0	0.430	0.168	0.100	0.058	0.017	0.004	0.001	0.000	0.000	0.000	0.000
	1	0.383	0.336	0.267	0.198	0.090	0.031	0.008	0.001	0.000	0.000	0.000
	2	0.149	0.294	0.311	0.296	0.209	0.109	0.041	0.010	0.004	0.001	0.000
	3	0.033	0.147	0.208	0.254	0.279	0.219	0.124	0.047	0.023	0.009	0.000
	4	0.005	0.046	0.087	0.136	0.232	0.273	0.232	0.136	0.087	0.046	0.005
	5	0.000	0.009	0.023	0.047	0.124	0.219	0.279	0.254	0.208	0.147	0.033
	6	0.000	0.001	0.004	0.010	0.041	0.109	0.209	0.296	0.311	0.294	0.149
	7	0.000	0.000	0.000	0.001	0.008	0.031	0.090	0.198	0.267	0.336	0.383
	8	0.000	0.000	0.000	0.000	0.001	0.004	0.017	0.058	0.100	0.168	0.430
9	0	0.387	0.134	0.075	0.040	0.010	0.002	0.000	0.000	0.000	0.000	0.000
	1	0.387	0.302	0.225	0.156	0.060	0.018	0.004	0.000	0.000	0.000	0.000
	2	0.172	0.302	0.300	0.267	0.161	0.070	0.021	0.004	0.001	0.000	0.000
	3	0.045	0.176	0.234	0.267	0.251	0.164	0.074	0.021	0.009	0.003	0.000
	4	0.007	0.066	0.117	0.172	0.251	0.246	0.167	0.074	0.039	0.017	0.001
	5	0.001	0.017	0.039	0.074	0.167	0.246	0.251	0.172	0.117	0.066	0.007
	6	0.000	0.003	0.009	0.021	0.074	0.164	0.251	0.267	0.234	0.176	0.045
	7	0.000	0.000	0.001	0.004	0.021	0.070	0.161	0.267	0.300	0.302	0.172
	8	0.000	0.000	0.000	0.000	0.004	0.018	0.060	0.156	0.225	0.302	0.387
	9	0.000	0.000	0.000	0.000	0.000	0.002	0.010	0.040	0.075	0.134	0.387
10	0	0.349	0.107	0.056	0.028	0.006	0.001	0.000	0.000	0.000	0.000	0.000
	1	0.387	0.268	0.188	0.121	0.040	0.010	0.002	0.000	0.000	0.000	0.000
	2	0.194	0.302	0.282	0.233	0.121	0.044	0.011	0.001	0.000	0.000	0.000
	3	0.057	0.201	0.250	0.267	0.215	0.117	0.042	0.009	0.003	0.001	0.000
	4	0.011	0.088	0.146	0.200	0.251	0.205	0.111	0.037	0.016	0.006	0.000
	5	0.001	0.026	0.058	0.103	0.201	0.246	0.201	0.103	0.058	0.026	0.001
	6	0.000	0.006	0.016	0.037	0.111	0.205	0.251	0.200	0.146	0.088	0.011
	7	0.000	0.001	0.003	0.009	0.042	0.117	0.215	0.267	0.250	0.201	0.057
	8	0.000	0.000	0.000	0.001	0.011	0.044	0.121	0.233	0.282	0.302	0.194
	9	0.000	0.000	0.000	0.000	0.002	0.010	0.040	0.121	0.188	0.268	0.387
	10	0.000	0.000	0.000	0.000	0.000	0.001	0.006	0.028	0.056	0.107	0.349
11	0	0.314	0.086	0.042	0.020	0.004	0.000	0.000	0.000	0.000	0.000	0.000
	1	0.384	0.236	0.155	0.093	0.027	0.005	0.001	0.000	0.000	0.000	0.000
	2	0.213	0.295	0.258	0.200	0.089	0.027	0.005	0.001	0.000	0.000	0.000
	3	0.071	0.221	0.258	0.257	0.177	0.081	0.023	0.004	0.001	0.000	0.000
	4	0.016	0.111	0.172	0.220	0.236	0.161	0.070	0.017	0.006	0.002	0.000
	5	0.002	0.039	0.080	0.132	0.221	0.226	0.147	0.057	0.027	0.010	0.000
	6	0.000	0.010	0.027	0.057	0.147	0.226	0.221	0.132	0.080	0.039	0.002
	7	0.000	0.002	0.006	0.017	0.070	0.161	0.236	0.220	0.172	0.111	0.016
	8	0.000	0.000	0.001	0.004	0.023	0.081	0.177	0.257	0.258	0.221	0.071
	9	0.000	0.000	0.000	0.001	0.005	0.027	0.089	0.200	0.258	0.295	0.213
	10	0.000	0.000	0.000	0.000	0.001	0.005	0.027	0.093	0.155	0.236	0.384
	11	0.000	0.000	0.000	0.000	0.000	0.000	0.004	0.020	0.042	0.086	0.314

Numbers in the table represent $p(X=x)$ for a binomial distribution with n trials and probability of success p.

Binomial probabilities:

$$\binom{n}{x} p^x(1-p)^{n-x}$$

n	x	0.1	0.2	0.25	0.3	0.4	0.5	0.6	0.7	0.75	0.8	0.9
12	0	0.282	0.069	0.032	0.014	0.002	0.000	0.000	0.000	0.000	0.000	0.000
	1	0.377	0.206	0.127	0.071	0.017	0.003	0.000	0.000	0.000	0.000	0.000
	2	0.230	0.283	0.232	0.168	0.064	0.016	0.002	0.000	0.000	0.000	0.000
	3	0.085	0.236	0.258	0.240	0.142	0.054	0.012	0.001	0.000	0.000	0.000
	4	0.021	0.133	0.194	0.231	0.213	0.121	0.042	0.008	0.002	0.001	0.000
	5	0.004	0.053	0.103	0.158	0.227	0.193	0.101	0.029	0.011	0.003	0.000
	6	0.000	0.016	0.040	0.079	0.177	0.226	0.177	0.079	0.040	0.016	0.000
	7	0.000	0.003	0.011	0.029	0.101	0.193	0.227	0.158	0.103	0.053	0.004
	8	0.000	0.001	0.002	0.008	0.042	0.121	0.213	0.231	0.194	0.133	0.021
	9	0.000	0.000	0.000	0.001	0.012	0.054	0.142	0.240	0.258	0.236	0.085
	10	0.000	0.000	0.000	0.000	0.002	0.016	0.064	0.168	0.232	0.283	0.230
	11	0.000	0.000	0.000	0.000	0.000	0.003	0.017	0.071	0.127	0.206	0.377
	12	0.000	0.000	0.000	0.000	0.000	0.000	0.002	0.014	0.032	0.069	0.282
13	0	0.254	0.055	0.024	0.010	0.001	0.000	0.000	0.000	0.000	0.000	0.000
	1	0.367	0.179	0.103	0.054	0.011	0.002	0.000	0.000	0.000	0.000	0.000
	2	0.245	0.268	0.206	0.139	0.045	0.010	0.001	0.000	0.000	0.000	0.000
	3	0.100	0.246	0.252	0.218	0.111	0.035	0.006	0.001	0.000	0.000	0.000
	4	0.028	0.154	0.210	0.234	0.184	0.087	0.024	0.003	0.001	0.000	0.000
	5	0.006	0.069	0.126	0.180	0.221	0.157	0.066	0.014	0.005	0.001	0.000
	6	0.001	0.023	0.056	0.103	0.197	0.209	0.131	0.044	0.019	0.006	0.000
	7	0.000	0.006	0.019	0.044	0.131	0.209	0.197	0.103	0.056	0.023	0.001
	8	0.000	0.001	0.005	0.014	0.066	0.157	0.221	0.180	0.126	0.069	0.006
	9	0.000	0.000	0.001	0.003	0.024	0.087	0.184	0.234	0.210	0.154	0.028
	10	0.000	0.000	0.000	0.001	0.006	0.035	0.111	0.218	0.252	0.246	0.100
	11	0.000	0.000	0.000	0.000	0.001	0.010	0.045	0.139	0.206	0.268	0.245
	12	0.000	0.000	0.000	0.000	0.000	0.002	0.011	0.054	0.103	0.179	0.367
	13	0.000	0.000	0.000	0.000	0.000	0.000	0.001	0.010	0.024	0.055	0.254
14	0	0.229	0.044	0.018	0.007	0.001	0.000	0.000	0.000	0.000	0.000	0.000
	1	0.356	0.154	0.083	0.041	0.007	0.001	0.000	0.000	0.000	0.000	0.000
	2	0.257	0.250	0.180	0.113	0.032	0.006	0.001	0.000	0.000	0.000	0.000
	3	0.114	0.250	0.240	0.194	0.085	0.022	0.003	0.000	0.000	0.000	0.000
	4	0.035	0.172	0.220	0.229	0.155	0.061	0.014	0.001	0.000	0.000	0.000
	5	0.008	0.086	0.147	0.196	0.207	0.122	0.041	0.007	0.002	0.000	0.000
	6	0.001	0.032	0.073	0.126	0.207	0.183	0.092	0.023	0.008	0.002	0.000
	7	0.000	0.009	0.028	0.062	0.157	0.209	0.157	0.062	0.028	0.009	0.000
	8	0.000	0.002	0.008	0.023	0.092	0.183	0.207	0.126	0.073	0.032	0.001
	9	0.000	0.000	0.002	0.007	0.041	0.122	0.207	0.196	0.147	0.086	0.008
	10	0.000	0.000	0.000	0.001	0.014	0.061	0.155	0.229	0.220	0.172	0.035
	11	0.000	0.000	0.000	0.000	0.003	0.022	0.085	0.194	0.240	0.250	0.114
	12	0.000	0.000	0.000	0.000	0.001	0.006	0.032	0.113	0.180	0.250	0.257
	13	0.000	0.000	0.000	0.000	0.000	0.001	0.007	0.041	0.083	0.154	0.356
	14	0.000	0.000	0.000	0.000	0.000	0.000	0.001	0.007	0.018	0.044	0.229

(continued)

Table A-3 (continued)

Numbers in the table represent $p(X=x)$ for a binomial distribution with n trials and probability of success p.

Binomial probabilities:

$$\binom{n}{x} p^x (1-p)^{n-x}$$

n	x	0.1	0.2	0.25	0.3	0.4	0.5	0.6	0.7	0.75	0.8	0.9
15	0	0.206	0.035	0.013	0.005	0.000	0.000	0.000	0.000	0.000	0.000	0.000
	1	0.343	0.132	0.067	0.031	0.005	0.000	0.000	0.000	0.000	0.000	0.000
	2	0.267	0.231	0.156	0.092	0.022	0.003	0.000	0.000	0.000	0.000	0.000
	3	0.129	0.250	0.225	0.170	0.063	0.014	0.002	0.000	0.000	0.000	0.000
	4	0.043	0.188	0.225	0.219	0.127	0.042	0.007	0.001	0.000	0.000	0.000
	5	0.010	0.103	0.165	0.206	0.186	0.092	0.024	0.003	0.001	0.000	0.000
	6	0.002	0.043	0.092	0.147	0.207	0.153	0.061	0.012	0.003	0.001	0.000
	7	0.000	0.014	0.039	0.081	0.177	0.196	0.118	0.035	0.013	0.003	0.000
	8	0.000	0.003	0.013	0.035	0.118	0.196	0.177	0.081	0.039	0.014	0.000
	9	0.000	0.001	0.003	0.012	0.061	0.153	0.207	0.147	0.092	0.043	0.002
	10	0.000	0.000	0.001	0.003	0.024	0.092	0.186	0.206	0.165	0.103	0.010
	11	0.000	0.000	0.000	0.001	0.007	0.042	0.127	0.219	0.225	0.188	0.043
	12	0.000	0.000	0.000	0.000	0.002	0.014	0.063	0.170	0.225	0.250	0.129
	13	0.000	0.000	0.000	0.000	0.000	0.003	0.022	0.092	0.156	0.231	0.267
	14	0.000	0.000	0.000	0.000	0.000	0.000	0.005	0.031	0.067	0.132	0.343
	15	0.000	0.000	0.000	0.000	0.000	0.000	0.000	0.005	0.013	0.035	0.206
20	0	0.122	0.012	0.003	0.001	0.000	0.000	0.000	0.000	0.000	0.000	0.000
	1	0.270	0.058	0.021	0.007	0.000	0.000	0.000	0.000	0.000	0.000	0.000
	2	0.285	0.137	0.067	0.028	0.003	0.000	0.000	0.000	0.000	0.000	0.000
	3	0.190	0.205	0.134	0.072	0.012	0.001	0.000	0.000	0.000	0.000	0.000
	4	0.090	0.218	0.190	0.130	0.035	0.005	0.000	0.000	0.000	0.000	0.000
	5	0.032	0.175	0.202	0.179	0.075	0.015	0.001	0.000	0.000	0.000	0.000
	6	0.009	0.109	0.169	0.192	0.124	0.037	0.005	0.000	0.000	0.000	0.000
	7	0.002	0.055	0.112	0.164	0.166	0.074	0.015	0.001	0.000	0.000	0.000
	8	0.000	0.022	0.061	0.114	0.180	0.120	0.035	0.004	0.001	0.000	0.000
	9	0.000	0.007	0.027	0.065	0.160	0.160	0.071	0.012	0.003	0.000	0.000
	10	0.000	0.002	0.010	0.031	0.117	0.176	0.117	0.031	0.010	0.002	0.000
	11	0.000	0.000	0.003	0.012	0.071	0.160	0.160	0.065	0.027	0.007	0.007
	12	0.000	0.000	0.001	0.004	0.035	0.120	0.180	0.114	0.061	0.022	0.000
	13	0.000	0.000	0.000	0.001	0.015	0.074	0.166	0.164	0.112	0.055	0.002
	14	0.000	0.000	0.000	0.000	0.005	0.037	0.124	0.192	0.169	0.109	0.009
	15	0.000	0.000	0.000	0.000	0.001	0.015	0.075	0.179	0.202	0.175	0.032
	16	0.000	0.000	0.000	0.000	0.000	0.005	0.035	0.130	0.190	0.218	0.090
	17	0.000	0.000	0.000	0.000	0.000	0.001	0.012	0.072	0.134	0.205	0.190
	18	0.000	0.000	0.000	0.000	0.000	0.000	0.003	0.028	0.067	0.137	0.285
	19	0.000	0.000	0.000	0.000	0.000	0.000	0.000	0.007	0.021	0.058	0.270
	20	0.000	0.000	0.000	0.000	0.000	0.000	0.000	0.001	0.003	0.012	0.122

Index

personal expenses, pie charts for, 92
picture form, expressing your exam
    question in, 342–343
pie charts
  described, 16, 92
  evaluating, 97
  of lotto revenue, 92–94
  misleading graphs, 320
  for ordering takeout, 94–95
  for personal expenses, 92
  for projecting age trends, 95–96
  tips for, 97
placebo, 57
placebo effect, 57, 265–266
planning and designing surveys, 250–254
polls
  analyzing data in, 259
  bias in, minimizing, 13
  carrying out, 256–259
  collecting data for, 256–257
  conclusions, drawing, 259–260
  cover letter for, 253
  errors in interpretation of survey results,
    259–260
  ethical issues, 253
  following up, 257–259
  impact of, 248–250
  interpreting results of, 259–260
  media and statistics, 27
  organizing data in, 259
  overview, 11–12, 58, 245–246
  planning and designing, 250–254
  process for, 250
  purpose of your survey, clarifying, 251
  questions on, formulating, 253–254
  response rate, 258
  sample selection, 254–256
  sources for, 246–247
  target population, 250–252
  timing for, 252
  topics for, 248
  type of, 252
pollsters, 246
population, 47–48

population mean
  confidence intervals for, 201–204
  described, 51, 72
  hypothesis tests, 228–229
population proportion, 175, 206–207
population standard deviation
  confidence intervals for population mean
    when you don't know, 203–204
  confidence intervals for population mean
    when you know, 202–203
  confidence intervals for the difference of
    two population means when you don't
    know, 210–211
  confidence intervals for the difference of
    two population means when you know,
    208–210
  standard error, 167–169
population variability, 201
predictions
  interpretation, 315
  overview, 292–293
probabilities
  sample mean, 173–174
  sample proportion, 177–178
  t-table, 160
probability distributions, 133–134
Probability For Dummies (Rumsey), 100
problems, understanding what the
    question is asking in test, 339–340
process for surveys, 250
projecting age trends, pie charts for, 95–96
projecting from sample to population,
    314–315
properties
  correlation, 286
  normal distribution, 144–145
purpose of your survey, clarifying, 251
p-values
  calculating, 222–223
  defining, 221
  guidelines for, 224
  overview, 61, 220
  test statistics and, 220–221